The Natural Geochemistry of Our Environment

David H. Speidel
Allen F. Agnew

Westview Special Studies in Natural Resources and Energy Management

The Natural Geochemistry of Our Environment

Also of Interest

The Impact of the Geosciences on Critical Energy Resources, edited by Creighton A. Burk and Charles L. Drake

† *Geomorphological Processes,* E. Derbyshire, K. J. Gregory, and J. R. Hails

* *Climate and the Environment: The Atmospheric Impact on Man,* John F. Griffiths

Weather Modification: Technology and Law, edited by Ray Jay Davis and Lewis O. Grant

† *Man and Environmental Processes: A Physical Geography Perspective,* edited by K. J. Gregory and D. E. Walling

Ecology and Environmental Management: A Geographical Perspective, C. C. Park

Land Aridization and Drought Control, Victor A. Kovda

† *Climate Change and Society: Consequences of Increasing Atmospheric Carbon Dioxide,* William W. Kellogg and Robert Schware

Arctic Pleistocene History and the Development of Submarine Permafrost, Michael E. Vigdorchik

Submarine Permafrost on the Alaskan Continental Shelf, Michael E. Vigdorchik

† Available in hardcover and paperback.
* Available in paperback only.

Westview Special Studies in Natural Resources and Energy Management

The Natural Geochemistry of Our Environment
David H. Speidel and Allen F. Agnew

As concern grows regarding environmental pollution, its effects, and its prevention, a better understanding of natural environmental processes—of the natural chemical behavior of the elements present on the surface of the earth—and how they affect the earth's ability to resist or fall prey to man-induced pollution is essential. In this book, the authors present what is known about these natural processes based on the concept of geochemical cycles: elements exist in sinks or reservoirs in knowable quantities and move from reservoir to reservoir in measurable fluxes. They stress that this movement is primarily controlled by the behavior of water, that materials moved in solution in streams, as sediment, or adsorbed on that sediment account for more than 90 percent of all geochemical movements.

Considering the major reservoirs—the atmosphere, rivers and oceans, and plants, animals, and humans—the authors evaluate the chemical composition and natural variations of each. The changes induced by human activity are at present relatively small when compared to natural variations, but, warn the authors, though the earth appears forgiving of human actions, some very real limits may soon be approached. In some situations we may feel free to use the earth's resources liberally, but in others judicious use is essential.

Dr. Speidel is professor of geochemistry in the Department of Earth and Environmental Sciences, Queens College, City University of New York, where he has also served as dean of the science faculty. He is a consultant to various government agencies and has been a visiting scholar at the Congressional Research Service, Library of Congress. **Dr. Agnew** is senior specialist in environmental policy (mining and resources) for the Congressional Research Service and has served as director of water research centers and professor of geology at Washington State University and at Indiana University, state geologist of South Dakota, and as geologist with the U.S. Geological Survey.

The Earth was made so various that the mind of desultory man, studious of change and pleased with novelty, might be indulged.

—William Cowper
The Task

The Natural Geochemistry of Our Environment

David H. Speidel
and Allen F. Agnew

Westview Press / Boulder, Colorado

Westview Special Studies in Natural Resources and Energy Management

All rights reserved. No part of this publication may be reproduced or transmitted in any form or by any means, electronic or mechanical, including photocopy, recording, or any information storage and retrieval system, without permission in writing from the publisher.

Copyright © 1982 by Westview Press, Inc.

Published in 1982 in the United States of America by
 Westview Press, Inc.
 5500 Central Avenue
 Boulder, Colorado 80301
 Frederick A. Praeger, President and Publisher

Library of Congress Cataloging in Publication Data
Speidel, David H., 1938–
  The natural geochemistry of our environment.
  (Westview special studies in natural resources and energy management)
  Includes bibliographies and index.
  1. Geochemistry. 2. Environmental chemistry. I. Agnew, Allen Francis, 1918– II. Title. III. Series.
QE515.S68   551.9   82-1948
ISBN 0-86531-110-2   AACR2

Printed and bound in the United States of America

Contents

List of Figures and Tables ix
Foreword, *George E. Brown, Jr.* xiii
Acknowledgments xv

1 Introduction 1

    The Elements of Our Concern 1
    Reservoirs of Our Concern 7
    References Cited 9

2 Water 13

    Introduction 13
    The World Water Budget 15
    Fluxes 23
    References Cited 34

3 Geochemical Movement 37

    Introduction 37
    Transport by Streams 37
    Ice Transport 52
    Transport by the Atmosphere 52
    Sorbtion 66
    References Cited 71

4 Soil Reservoir 81

    Introduction 81
    Soil Formation 83
    Soil Composition 91
    Mobility of Elements 95
    References Cited 100

5 Ocean Reservoir 103
Introduction 103
Nutrient Elements and Chemical Reactions 111
Particulate Matter in the Oceans 115
Ocean Sediments 122
References Cited 128

6 Biota and the Biosphere 131
Introduction 131
Composition 134
Productivity and Biomass 156
Major Element Cycles 169
Trace Elements 176
References Cited 185

7 Geochemical Cycles and Fluxes Revisited: Summary 193
Water Reservoirs and Fluxes 193
Transport by Water 194
Transport by Air 195
References Cited 197

Epilogue: A Moral Dilemma 199
References Cited 200

Index 201

Figures and Tables

Figures

1-1   Concentration patterns of trace elements 4
1-2   Composition of the crust of the Earth 6
1-3   Geochemical reservoirs and fluxes 7
2-1   Major river drainage basins of the world (map) 28
2-2   Regions of surplus and deficit of river water resources (map) 33
3-1   Relationships of variations in dissolved load of streams
      with intensity of stream discharge 39
3-2   Mechanisms controlling salinity and chemical composition
      of world surface water; variation of salinity and
      $Cl/(Cl + HCO_3)$ 42
3-3   Relationship of variations in dissolved load of streams
      with solid load of streams 46
3-4   Effect of size of drainage area on sediment-delivery ratio 47
3-5   Change in concentration of suspended matter in different waters . . . 50
3-6   General distribution of soil erosion in the
      United States (maps) 56
3-7   Distribution of acid rain in northeastern United States
      in 1955–1956 and 1972–1973 (maps) 63
3-8   Buffering patterns of natural waters 64
3-9   pH changes in geochemical movement processes 65
3-10 Transport of some transition elements in the Amazon
      and Yukon rivers 68
3-11 Generalized model for cation and anion adsorption 70
4-1   The soil reservoir in the geochemical cycle 82
4-2   Land-surface classification in relation to processes that
      cause changes in landforms 86
4-3   Organic matter production in humid climates compared to
      that destroyed in aerated and unaerated humid climates 89
4-4   Soil-zone formation as a function of temperature
      and precipitation 89

4-5	Six broad soil zones of the world (map)	92
4-6	Ranges of measured abundance of elements in soils	93
4-7	Variation in major cation composition of soils with changing acidity	94
5-1	The ocean reservoir in the geochemical cycle	104
5-2	The concentration of selected elements in average sea water	105
5-3	Forms of occurrence of metal species	109
5-4	Typical oceanic concentration of nutrient elements with depth	112
5-5	Distribution of sediments on the ocean floor (map)	116
5-6	The silica cycle in the oceans	118
5-7	Factors controlling the accumulation of sediment under the equatorial high-productivity belt	119
5-8	Ocean circulation	121
5-9	A classification of the hydrogenous material in marine sediments	123
6-1	The biota reservoir in the geochemical cycle	132
6-2	Relative weights of some geochemical reservoirs	133
6-3	Composition of extracellular and intercellular fluids	135
6-4	Major cation compositional variation with biota type	140
6-5	Plant/soil enrichment ratios of elements as a function of ionic potential	143
6-6	Changes in normalized enrichment ratios for selected elements:	
	IA—Li, Na, K, Rb	146
	IIA—Be, Mg, Ca, Sr, Ba	147
	IIB—Zn, Cd, Hg	148
	IIIA—B, Al, La	149
	IV—C, Si, Ge, Sn, Pb, Ti, Zr	150
	Metals—Cr, Mn, Fe, Co, Ni, Cu, Mo	151
	VB—N, P, As, Sb	152
	VIB—O, S, Se	153
	VIIB—F, Cl, Br, I	154
6-7	Distribution of major biomes as a function of temperature and precipitation	157
6-8	Distribution of generalized biome types (map)	158
6-9	Distribution of dry organic material in selected world biomes	164
6-10	Annual biogeochemical cycle of elements in a 117-year-old oak-hazel forest in Belgium	166
6-11	Comparison of the annual biogeochemical cycle of nutrient elements in different types of terrestrial ecosystems	168

Figures and Tables   xi

6-12   Carbon reservoirs and fluxes in the geochemical cycle 170
6-13   Sulfur reservoirs and fluxes in the geochemical cycle 172
6-14   Nitrogen reservoirs and fluxes in the geochemical cycle 173
6-15   Phosphorus reservoirs and fluxes in the geochemical cycle 175
6-16a Areas in the United States where trace-element deficiencies
       occur in crops (map) 178
6-16b Areas in the United States where trace-element–related
       diseases (toxicity) occur in animals (map) 179

Tables

1-1	Elements of our concern	3
2-1	Water use of the world and the United States	14
2-2	Water reservoirs and fluxes	16
2-3	Geographic distribution of the world water cycle	17
2-4	Distribution of continental runoff: source of runoff to oceans	25
2-5	General features of the world's biggest rivers	26
2-6	Oceanic water residence times	30
3-1	Geochemical movement	38
3-2	Dissolved load discharged to world ocean	40
3-3	Composition of river water, worldwide average for major ions in solution	41
3-4	Comparison of water chemistry	44
3-5	Natural production of atmospheric particles	54
3-6	Atmospheric trace element enrichment coefficient	58
3-7	Elements adsorbed on iron and manganese oxides on soil particles and stream sediments	71
4-1	Chemical elements depleted and concentrated in soil	84
4-2	Land form units and processes	87
4-3	Mode of occurrence of trace elements in soil and stream sediments	96
4-4	Relative mobility of elements with variation in acidity and oxidizing conditions	98
5-1	Major ion speciation in sea water	107
5-2	Range of estimates of speciation of selected trace elements in sea water	108
5-3	Concentrations of some trace elements in deep-sea sediments	124
5-4	Elements adsorbed on marine Fe-Mn nodules	127
6-1	Major element atomic ratios in biota	136
6-2	Soil, crust, and land-plant element compositions	144

6-3	Net primary productivity, biomass, and biomass distribution of major vegetation units of the world	160
6-4	Known or suspected effects of anomalous levels of trace elements on plants and animals, including humans	180
6-5	Composition of U.S. coals	184

Foreword

A rational and responsive use of the environment by U.S. citizens has been a concern of mine for many years. Along with many of my colleagues in the U.S. Congress who share this concern, I have attempted to oversee a number of federal executive regulatory agencies in a meaningful manner. But, to fulfill this responsibility, I must know the mechanisms of the natural environment and the potential for their being modified in various ways and degrees. This means that I must have a broad understanding of natural environmental processes.

This book, which is an outgrowth of a report prepared in 1979 by Drs. Speidel and Agnew for me when I was chairman of the Subcommittee on Science, Research, and Technology, provides just such a background. It enables one to comprehend the natural system and the way that human activities affect that environment.

The Natural Geochemistry of Our Environment shows that the Earth is a water world, whose water is transformed readily from the solid to the liquid to the gaseous state. The time scale for such transfers, called fluxes, ranges from a few weeks to thousands of years. These fluxes are dominated by precipitation and evaporation; differences in these processes are caused by the land-ocean relationships, the rotation of the Earth, and by the intensity of radiation that reaches the Earth from the Sun.

The water flux accounts for most (98 percent) of the movement of the chemical elements from one Earth reservoir to another—both as physical particles and as chemically dissolved matter. The remaining 2 percent is transported by air. This book by Drs. Speidel and Agnew attempts to document those movements, evaluate the reliability of the numbers presented therein, and—most importantly—show the limits of our knowledge.

This book provides a background for considering the environmental dilemmas confronting society today. It tells us that we must accommodate our lives to the natural geochemical cycles of the Earth, recognizing that in some situations we may use the Earth's resources freely with little or no harm as a result, but that in all other situations we must be far more judicious in their use than we have been in the past.

The authors are to be congratulated for calling our attention to the importance of the baseline structure and mechanism of our environment—perhaps more important than many of the "crises" that so often command greater attention. If we cannot strengthen our understanding of that baseline we will be at grave risk of being blind to even profoundly dangerous changes.

George E. Brown, Jr.
House of Representatives
U.S. Congress

Acknowledgments

This study was initiated when one of us (DHS) spent a sabbatical year as a visiting scholar with the Senior Specialists' Division, Congressional Research Service (CRS), Library of Congress. We thank Mr. Gilbert Gude, director of CRS, and Dr. John Hardt, chief of the Senior Specialists' Division, for making that possible. This work was originally just one section of a much larger work on biogeochemical research in the United States directed by James McCullough, now chief, Science Policy Research, CRS. We thank him and his staff for their aid and encouragement. Congressman Ray Thornton (1977–1979) and Congressman George E. Brown, Jr. (1979–1981) were chairmen of the Science, Research, and Technology subcommittee of the Committee on Science and Technology that requested the study. Gail Pesyna, T. R. Kramer, and P. B. Yeager, staff members of the committee, coordinated the study. L. Cinquemani, M. Rafanelli, S. Sachnoff, and J. Taylor of Queens College aided in preparation of the final manuscript, as did M. Speidel. Our thanks to all of them.

D.H.S.
A.F.A.

1
Introduction

All things flow.
—Heraclitus

The agronomist, the ecologist, the hydrologist, the nutritionist, the geologist, the chemist, the oceanographer, and the atmospheric scientist—all, at one time or another—are interested in the behavior of chemical elements on the surface of the Earth. Clearly, they are not all interested in the same elements nor are they interested in the same portion of the Earth. The growth in scientific literature in quantity, sophistication, and specialization has made it increasingly difficult for a scientist in one area to gain a thorough appreciation for what is happening in other areas. Many detailed books exist for each of the chemical elements and for each of the disciplines. But the world is not compartmentalized into neat isolated academic boxes. Elements move. The result is that the chemical behavior of elements in any one environment present at the surface of the Earth cannot be fully known until it is known in all of the different environments. This book is an attempt to provide an overview of the Earth that will enable earth scientists to gain an appreciation of how and where the work of scientists in other disciplines can be applied to earth-science problems. It also will provide an overview of the natural geochemical behavior of elements for the scientists in these other disciplines so that they, in turn, can use the work of earth scientists for their own problems. We have attempted to achieve a balance between detail and sketches, but we may not have been successful throughout this book. We hope the reader will find that we have not weighted either side unbearably.

The Elements of Our Concern

Geologists are usually concerned with elements that make up the largest fraction of rocks and fresh water. Ecologists, on the other hand, are most concerned with those elements that are critical to plants or animals (1). Unfortunately, geologists' elements tend not to be the same as those of ecologists; this can account for many of the differences in interpretation by government regulators of the amounts of particular elements that should require regulation. In this book we shall focus on those elements that are of biological importance (Table 1-1) — because the origins of life and the functioning of life processes can thus be shown to be a function of geochemical and related environmental factors. This area of interest has been termed *geochemical ecology* (8). It includes the "dependence of life and the formation of living matter on geochemical and related environmental factors, the migration paths of chemical elements in the biosphere, the biochemical fixation of these elements by organisms, and their incorporation in living organisms during metabolic processes, biological reactions caused by an excess or a deficiency of chemical elements" and the determination of numerical values that cause such reactions.

The major elements are the principal components of living matter — the carbohydrate building block, CH_2O, and the macronutrients N, Mg, P, S, K, and Ca. Note in Table 1-1 that Na and Cl, although not major elements in plants, are important in animals, where they produce gastric secretion, regulate acid-base balance, and promote the proper fluid balance. The micronutrients, or essential trace elements, are also listed in Table 1-1. The names given to the different categories vary with the discipline involved. Nutritionists, for example, refer to minerals (Ca, P, Mg, Fe, Zn, and I), trace elements (Cu, Mn, F, Cr, Se, and Mo), and electrolytes (Na, K, and Cl) (7). We use the term "essential" to signify that experimentation has documented abnormal pathology either in growth, function, or behavior in a particular organism when that element is withheld from, or is severely limited in availability to, the organism. With many elements, the function they serve is not always known, but it is known that their absence results in an abnormality. Some elements, such as boron (B), are needed for plants but they do not appear to be essential for animals. The increased biological complexity of animals as compared with plants is indicated by animal dependence on a larger number of chemical elements — generally those that serve as catalysts or have a rate-controlling function. Note that few elements with large atomic numbers (greater than 34, which selenium has) appear to be essential — only molybdenum (Mo), iodine (I), and tin (Sn). Elements that appear to be important neither as trace nutrients nor as toxic elements are not considered to any significant extent in later chapters.

The breakdown of elements into essential and nonessential is not clear cut and the confusion regarding toxic elements is even greater. In Figure 1-1 we have considered some patterns that might exist for essential and nonessential elements (2, 3, 4, 9). If a trace element is essential, too little of it will create an inactive zone — a dose response that is less than what is considered "normal" for the organism. There will be a range of concentrations that corresponds to a region of optimal supplementation for normal function of the organism. This

TABLE 1-1
Elements of Our Concern

Elements Necessary for Plants		Elements Necessary for Animals		Elements That Are Nonessential
Major	Micronutrients	Major	Micronutrients	
Hydrogen, H	Boron, B	Hydrogen, H	Fluorine, F	Lithium, Li
Carbon, C	Chlorine, Cl	Carbon, C	Silicon, Si	Beryllium, Be
Nitrogen, N	Manganese, Mn	Nitrogen, N	Vanadium, V	Boron, B
Oxygen, O	Iron, Fe	Oxygen, O	Chromium, Cr	Aluminum, Al
Magnesium, Mg	Copper, Cu	Sodium, Na	Manganese, Mn	Titanium, Tl
Phosphorus, P	Zinc, Zn	Magnesium, Mg	Iron, Fe	Germanium, Ge
Sulphur, S	Molybdenum, Mo	Phosphorus, P	Cobalt, Co	Bromine, Br
Potassium, K		Sulphur, S	Nickel, N	Rubidium, Rb
Calcium, Ca		Chlorine, Cl	Copper, Cu	Strontium, Sr
		Potassium, K	Zinc, Zn	Zirconium, Zr
		Calcium, Ca	Arsenic, As	Silver, Ag
			Selenium, Se	Cadmium, Cd
			Molybdenum, Mo	Antimony, Sb
			Tin, Sn	Tellurium, Te
			Iodine, I	Barium, Ba
				Rare Earth, R.E.
				Mercury, Hg
				Lead, Pb
				Thorium, Th
				Uranium, U

Source: Refs. 2-6 and others. The elements not considered in the present study are: scantium, gallium, yttrium, niobium, technicium, ruthenium, rhodium, palladium, indium, cesium, hafnium, tantalum, tungsten, rhenium, osmium, iridium, platinum, gold, thallium, bismuth, polonium, astatine, francium, radon, actinium, and the noble gases. In addition to the recommended daily dietary allowances for Cu and Mo mentioned in the text, the following are also recommended (7) for adult humans: Ca, 800 mg; P, 800 mg; Mg, 300 mg; Fe, 15 mg; Zn, 15 mg; I, 150 mg; Mn, 2.5-5 mg; F, 1.5-4 mg; Cr, 0.05-0.2 mg; Se, 0.05-0.2 mg; Na, 1100-3300 mg; K, 1875-5625 mg; and Cl, 1700-5100 mg.

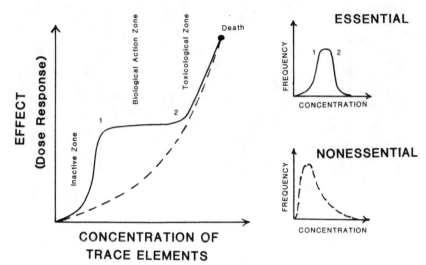

FIGURE 1-1. Concentration patterns of trace elements. Patterns for essential elements are given by solid lines. Patterns for nonessential elements are given by dashed lines. Threshold values are not indicated. Sources: Refs. 2, 3, 4, 8, 9.

range, indicated by points 1 and 2 in Figure 1-1, responds to particular capacity and tolerance limits of the animal or system. It also responds to other elements and compounds present in the body and diet (10). A further increase in concentration will have an adverse effect, causing toxicological action and, eventually, death.

The toxic levels for many trace elements may be only several times greater than the concentrations found for the normal range. Once the homeostatic capacity (indicated by the range of concentrations between points 1 and 2 in Figure 1-1) is exceeded, slight changes in concentration can have major toxicological effects. Thus, normal organisms are expected to have concentrations that range between points 1 and 2; below 1 organisms cannot function, and above 2 they have toxic reactions. Such a distribution should give the normal curve indicated on the upper right of Figure 1-1 (2, 9, 11). Examples of such ranges (2) include:

- Cu
  - less than 15 ppm causes anemia, bone diseases, and nonripening of cereals;
  - 15–60 ppm is normal;
  - greater than 60 ppm causes plant chlorosis.
- Mo
  - less than 1.5 ppm causes plant diseases;
  - 1.5–4 ppm is normal;
  - greater than 4 ppm causes goiter and Mo toxicity.

The Food and Nutrition Board of the National Academy of Sciences indicates that the normal amount of Cu in an adult human can be maintained by ingesting 2-3 mg daily and for Mo, by ingesting 0.15-0.5 mg (7).

Nonessential elements will not show a range that corresponds to the biological action zone indicated in Figure 1-1. Instead, they will follow a highly skewed distribution with the most frequent concentrations of little or no biological effect. For example:

Sr
- up to 600 ppm is normal;
- greater than 600 ppm can result in brittleness of bones and arthritis.

In drawing the dashed curve of Figure 1-1 for nonessential elements, we did not indicate a threshold — that is, a certain concentration that must be reached before any dose response is found. The examples given above would indicate that, at least in the case of strontium, a threshold is a reasonable speculation.

If this is the type of argument used to demonstrate the essential nature of a particular element, it is clear that the list is not fixed. Elements will be added (and possibly subtracted) as more analyses are done and as analytical accuracy is improved. Indeed, one major problem referred to throughout this book is that of analytical accuracy. Turekian recently pointed out that trace-element geochemistry would not have been nearly as exciting to study were it not for the disputes based on analytical inaccuracies caused by the immense difficulties of performing the analysis (12). There is a major problem with low-number statistics: Lack of reproducibility is a function of sample size and that discrepancy can approach a factor of two if the sample volumes differ by a factor of five (13). This is, unfortunately, not uncommon in analyses of ocean, atmosphere, and biosphere. Thus, the numbers presented in this book are "fuzzy" and significant differences between today's explanation of the behavior of elements and that in the future can be expected to develop as our analyses become more accurate. "Values are approximations, computed on data from many sources which are not mutually consistent. None of the values is precise" (14).

One last point on the elements is of interest. The composition of the crust of the Earth provides an interesting comparison with the biologically important elements. The range of values for particular elements presented in recent tabulations (15-25) is indicated by the length of the bars in Figure 1-2. Major elements of Table 1-1 are indicated by the largest letters, essential trace elements by the next size, and toxic or nonessential trace elements by the smallest size. The relative abundances of those elements not discussed in this book are indicated by the positions of their chemical symbols on the right side of the figure. Part of the variation in composition values is because individual analyses are not made on the mythical average crustal rock. Each rock type

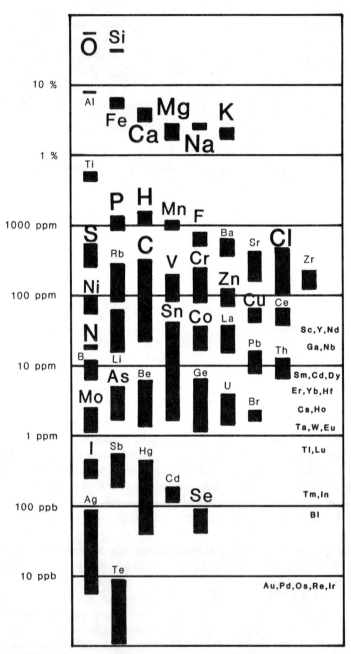

FIGURE 1-2. Composition of the crust of the Earth. For explanation, see text. Sources: Refs. 15 through 25.

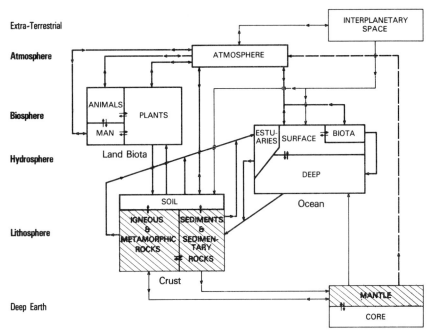

FIGURE 1-3. Geochemical reservoirs and fluxes. Heavy lines indicate fluxes examined. Heavy outlines on boxes indicate reservoirs of special interest in exogenic processes. Shaded areas indicate major sources and sinks for elements in the cycle.

has to be weighted to get an average. Green, for example, tried to avoid this problem by tabulating averages for 16 separate geospheres (18). Throughout this book we will face the problem of correct weighting of each subgroup to get the composition of the whole. The range of values for a particular element shown in Figure 1-2 indicates the uncertainty to be expected.

Reservoirs of Our Concern

Having stated previously that the world does not exist in neat boxes, we will now attempt to construct some. These boxes, or reservoirs, provide a convenient focus for our attention. We will assume that the chemical elements exist in these reservoirs in determinable quantities and move from reservoir to reservoir in measurable fluxes. The behavior of the element is a function of its inherent chemical attributes, its interaction with environmental factors, and the presence of other elements. The major reservoirs are indicated in Figure 1-3. Although they are simplifications, they represent those that can be described, developed, and determined by small-scale investigation. The reservoirs of interest in this book, which are indicated by a heavier outline, have some available data on their compositions and perform major functions in the biosphere—that is, they form the exogenic (or surface) cycle of earth processes. Any complete study of geochemical cycles should include the addition and loss of material from outside the Earth (extraterrestrial processes) or from within (mantle-core processes). Unless specifically mentioned, such changes are assumed herein to be negligible. The fact that not all reservoirs interact with each other allows us to simplify our discussion of them.

Questions of scale cause major problems in perception. How much water is in the ocean? The atmosphere? Glaciers? How much SiO_2 is present in the crust of the earth? The sediments of the ocean? The atmosphere? What is the significance of such huge numbers? Do they have meaning in the aggregate or do only their constituent parts have meaning?

In evaluating both reservoirs and fluxes, two parameters must be known: (a) the chemical composition of each of the component parts and (b) the volume of each of those parts present in the reservoir or flux. For example, to know the composition of the human body, one must know the concentration of C, N, O, Fe, Cu, Co, etc. in the heart, kidneys, bones, muscles, etc., and must know the proportion of the weight of those body parts to the weight of the total human body. Other examples are the determination of the composition of trees, grasses, and animals, and their amounts, so as to calculate the total composition of the biosphere; or the composition and flow rate of all rivers in the world, so as to determine the composition and quantity of dissolved constituents moved by the flux of surface water.

These evaluation examples also point out some of the major problems inherent in the processes. Even if there is no analytical problem, what constitutes a "normal" heart? Whose human body are we talking about? How do we calculate how many sugar maples there are in the world? How do we compensate for the effects of floods on individual streams? How does one deal with the normal variations of composition of the component parts and variation in amount of those parts of the phases? One can integrate and smooth such variation over many streams and many years, over many organs and all types of people; such averages may be useful in large-scale, long-term calculations, but may have no validity for establishing the "normal" or instant behavior against which the variation of a particular body or a particular stream can be measured.

Ranges of values can be generated that give the extreme values for normal (nonpathological) behavior, and specific examples can be compared to such ranges. For example, although the normal range of iron in the human body appears to be 60–70 ppm (2), dietary variation can cause individuals to move outside that range without reaction or, conversely, inside that range with some reaction. Thus, the individual case must be examined. Variation in concentration of SiO_2 in streams of the Amazon basin is different from SiO_2 behavior in the Potomac and Susquehanna basins. Indeed, the composition of streams and the quantity of water in them vary greatly with season.

Residence time (see Chapter 2) is a measure of the movement of a particular element through a reservoir. On a geological time scale, the assumptions are usually made that a steady state exists (quantity in equals quantity out) and/or that the actual amount moved is a function of the amount present (26-33). On a localized time scale, looping (34) can add a severe disruption to the analytical framework. In these cases, the differential movement between endpoints is always less than the total of impact plus transfer because the impact items bounce and are counted several times. The problem of how to model the fluxes is a severe one and several recent attempts (28, 29, 31) are especially worth examining.

All of the geobiochemical processes that will be discussed in this book are "exogenic" processes on Earth, and have as the prime energy source the dissipation of electromagnetic radiation that arrives at the Earth's surface and heats its atmosphere. The dominant processes are chemical weathering, wind or water erosion and transportation, deposition, chemical precipitation, and rock formation. It is interesting to note that Venus and the Earth have the same energy source, and the processes on both planets are expected to be similar. However, on planets or satellites with no atmosphere, such as Mercury or the Moon, the major energy source is by direct transmission in the form of meteorites, radiation, and impact of atomic and nuclear particles. The Apollo mission has emphasized the differences in surface form of such planetary bodies that result from the different processes.

The Earth's unique aspect is caused by the presence of water in liquid, solid, and vapor form. In succeeding chapters we shall examine the behavior of the water flux through the Earth's reservoirs.

References Cited

1. Vitousek, P. M. and W. A. Reiners, 1976, Ecosystem Development and the Biological Control of Stream Water Chemistry; p. 665-680 *in* Environmental Biogeochemistry, v. 2, J. O. Nriagu, ed., Ann Arbor Science, Mich., 797 p.

2. Underwood, E. J., 1977, Trace Elements in Human and Animal Nutrition; 4th ed., Academic Press, New York, 545 p.

3. Bowen, H.J.M., 1966, Trace Elements in Biochemistry; Academic Press, London, 241 p.

4. Bowen, H.J.M., 1979, Environmental Chemistry of the Elements, Academic Press, New York, 334 p.

5. Frieden, E., 1972, The Chemical Elements of Life; Scientific American, v. 227, no. 1, p. 52-60.

6. Brooks, R. R., 1972, Geobotany and Biogeochemistry in Mineral Exploration; Harper and Row, New York, 290 p.

7. National Academy of Sciences, 1980, Recommended Daily Dietary Allowances; Food and Nutrition Board, NAS-NRC, Washington, D.C.

8. Koval'skii, V. V., 1975, Some Tasks of Geochemical Ecology; p. 575-582 in Recent Contributions to Geochemistry and Analytical Chemistry, A. I. Tugainov, ed., Halstead Press, New York (original Russian in 1972).

9. Liebscher, K. and H. Smith, 1968, Essential and Non-essential Trace Elements; Archives of Environmental Health, v. 17, p. 881-890.

10. Levander, O. A. and L. Cheng, 1980, Micronutrient Interactions: Vitamins, Minerals, and Hazardous Elements; Annals of the New York Academy of Sciences, v. 355, New York, 372 p.

11. Tipton, I. H. and M. J. Cook, 1963, Trace Elements in Human Tissue II; Health Physics, v. 9, p. 103-145.

12. Turekian, K. T., 1977, The Fate of Metals in the Oceans; Geochimica et Cosmochimica Acta, v. 41, p. 1139-1144.

13. Michels, D. E., 1977, Sample Size Effect on Geometric Average Concentrations for Log-Normally Distributed Contaminants; Environmental Science and Technology, v. 11, no. 3, p. 300−302.

14. Nace, R. L., 1969, World Water Inventory and Control; p. 31-42 in Water, Earth, and Man, R. J. Chorley, ed., Methuen, London, 588 p.

15. Goldschmidt, V. M., 1954, Geochemistry; Clarendon Press, Oxford, 730 p.

16. Ahrens, L. H., 1965, Distribution of Elements in our Planet; McGraw-Hill, New York, 110 p.

17. Beus, A. A. and S. V. Grigorian, 1975, Geochemical Exploration Methods for Mineral Deposits; Applied Publishing Ltd., Wilmette, Ill., 287 p.

18. Green, J., 1972, Elements: Planetary Abundances and Distribution; p. 268-300 in Encyclopedia of Geochemistry and Environmental Science, R. Fairbridge, ed., Van Nostrand Reinhold, New York, 1321 p.

19. Mason, B., 1966, Principles of Geochemistry; 3rd ed., Wiley, New York, 329 p.

20. Parker, R. L., 1967, Composition of the Earth's Crust; U.S. Geological Survey Professional Paper 440-D, Washington, D.C., 17 p.

21. Perel'man, A. I., 1967, Geochemistry of Epigenesis; trans. by N. N. Kohanowski, Plenum, New York, 266 p.

22. Ronov, A. B. and A. A. Yaroshevsky, 1969, Chemical Composition of the Earth's Crust; p. 37-57 in The Earth's Crust and Upper Mantle, P. Hart, ed., American Geophysical Union, Washington, D.C.

23. Ronov, A. B. and A. A. Yaroshevsky, 1972, Earth's Crust Geochemistry; p. 243-254 in Encyclopedia of Geochemistry and Environmental Science, R. Fairbridge, ed., Van Nostrand Reinhold, New York, 1321 p.

24. Rosler, H. J. and H. Lange, 1972, Geochemical Tables; Elsevier, New York.

25. Taylor, S. R., 1964, Abundance of Chemical Elements in the Continental Crust: A New Table; Geochimica et Cosmochimica Acta, v. 28, p. 1273-1285.

26. Bolin, B., 1976, Transfer Processes and Time Scales in Biogeochemical Cycles; p. 17-22 in Nitrogen, Phosphorus and Sulfur-Global Cycles, SCOPE Report 7, Ecology Bulletin, v. 22, Stockholm, 192 p.

27. Bolin, B. and H. Rodhe, 1973, A Note on the Concepts of Age Distribution and Transit Time in Natural Reservoirs; Tellus, v. 25, no. 1, p. 58-62.

28. Holland, H. D., 1978, The Chemistry of the Atmosphere and Oceans; Wiley, New York, 351 p.

29. Lerman, A., 1979, Geochemical Processes, Water and Sediment Environments; Wiley, New York, 481 p. (See especially Chapters 1-3.)

30. Li, Yuan-Hui, 1972, Geochemical Mass Balance Among Lithosphere, Hydrosphere, and Atmosphere; American Journal of Science, v. 272, p. 119-137.

31. Mackenzie, F. T. and R. Wollast, 1977, Sedimentary Cycling Models of Global Processes; p. 739-785 *in* The Sea, v. 6, Marine Modeling, E. D. Goldberg, I. N. McCave, J. J. O'Brien, and J. H. Steele, eds., Wiley, New York, 1048 p.

32. Martin, B., 1976, Critical Evaluation of Residence Times Calculated Using the Exponential Approximation; Journal of Geophysical Research, v. 81, no. 15, p. 2637-2640.

33. Walker, J.C.G., 1977, Evolution of the Atmosphere; Macmillan, New York.

34. Odén, S., 1976, The Acidity Problem—An Outline of Concepts; Water, Air, and Soil Pollution, v. 6, p. 137–166.

2
Water

The noblest of elements is water.
—Pindar

Introduction

Water circulates from the ocean to the air to the land and back to the ocean, with some passing through the ground along the way. It is one of the controlling influences on climate and weather. Water is critical in the formation and stability of soil, playing the key role in erosion and transportation of sediment. As a resource, water is abundant and usually inexpensive. We drink it. We use it for bathing, for cooling and heating, for carrying away waste, for growing food, and for transporting goods. No wonder there are conflicts! However, until recently, we have become concerned about water only when it is no longer available to us either by its absence, by its degradation due to pollution, or when present in overabundance, as in floods.

From outer space, the Earth appears to have water as its most abundant substance. Almost 71 percent of our planet's surface is blanketed by oceans, and another 3.5 percent by polar ice. The land is covered with patches and linear streaks of clouds, snow, lakes, and rivers. The Earth is the only planet in our solar system where water is known to exist in its three forms, vapor, ice, and liquid water. The changes from one form to another create the opportunity for life on Earth.

The overall hydrologic cycle is easy to describe. A natural distillation process, powered by solar energy, creates evaporation at water surfaces and transpiration at plant surfaces; the resulting water vapor is transported in the atmosphere and returns to the Earth's surface as precipitation. The part of the precipitation that falls on the ocean begins the cycle again; the part that falls on the land can infiltrate the ground, accumulate on the surface, or move on the surface as runoff to streams and eventually to the sea.

Surface accumulations ultimately evaporate, returning the water as vapor to the atmosphere. Infiltrating water enters openings in the soil or rock until the pores are saturated. This reservoir of accumulated ground water slowly

TABLE 2-1
Water Use of the World and the United States (billion gallons per day)

	Water Withdrawn			Water Consumed		
	1962	1970	1975	1962	1970	1975
Irrigation						
U.S.	142	135	145	85	77	83
World	-	1919	-	-	1340	-
Industry						
U.S.	142	217	234	2.5	6	6
World	-	145	-	-	29	-
Municipal						
U.S.	24	27	29	2.5	6	7
World	-	72	-	-	41	-

Source: U.S. information from Refs. 1, 2, 3, 4 (Fig. 5-6 and Tables 5-73 and 5-74) and 5 (Tables 1-1 and 1-2). World information from Ref. 6. Ref. 7 has comparable values. Stream flow in the United States in 1975 was 1200 bgd of which 1105 reached the ocean. The units most often used internationally for water flow rate are km^3/yr. Irrigation water in North America is often measured in acre-feet - amount of water needed to cover 1 acre to a depth of 1 ft. The conversions are 1 km^3/yr = 811,000 acre-ft/yr = 723 mgd.

feeds springs and streams. Part of the infiltrating water is taken up by plant roots and, moving through the plant, is eventually released to the atmosphere from the leaves through transpiration in the process of photosynthesis. Runoff carries the dissolved and detrital results of rock weathering, creating possible problems both where the erosion takes place and where the resulting materials are deposited. The many millions of years of geological history can be followed through the varying cycles of erosion, transportation, deposition, lithification, uplift, and erosion again.

Human uses of water follow closely the consumptive and nonconsumptive patterns of nature. Boiling and irrigation uses are consumptive and are equivalent to natural evaporation. Washing, cooking, and processing uses are nonconsumptive, as is the carrying of water by streams to the rivers and the ocean. Dilution has been the major treatment of wastes in geological cycles and has utilized the same technique.

The amount of streamflow from runoff, about 30 percent of the U.S. precipitation, corresponds to more than 1200 billion gallons per day for the continental United States, as shown in Table 2-1. This quantity is reduced by withdrawals of water by humans for uses that have varying degrees of consumption. The consumption of agricultural water is equivalent to 140 mm

(5 in.) of water spread over an area the size of Texas and equivalent to the entire rainfall for the state of Arizona. Industrial use of water in 1975 increased both the amount and percentage of water withdrawn compared to 1972, but consumed only 7 percent of that withdrawn in 1975. Municipal water utilities withdraw only 2 percent of the streamflow, but this public water use consumed 23 percent of that amount, up sharply from 10 percent in 1962.

Thus, we see that approximately 35 percent of the runoff is withdrawn, with 23 percent being recycled into the streams and the remaining 12 percent transmitted directly back to the atmosphere. The recycled water can be used several times, which adds up to more than 100 percent. The addition of pollutants during water use creates many of the problems discussed later. These effects and their importance causes us to examine the water cycle in much closer detail in the following section.

The World Water Budget

Oceanic Water

The surface of the Earth is dominated by water, the exact amount of which is difficult to measure. The range of recent estimates of these amounts is presented in Table 2-2. There is tremendous difficulty in determining the volume of the ocean basins and in estimating the amount of ground water. There is only one world ocean, so any subdivisions into individual oceans is arbitrary. Generally, the divisions are the Atlantic, Pacific, Indian, and Arctic (or North Polar) oceans. There is general agreement about the total area covered by water but there is not always agreement as to which ocean a particular portion belongs to. For example, does the Arctic Ocean cover 8.509×10^6 km² (see Ref. 8) or 12.250×10^6 km² (see Ref. 9)? Because ocean volumes are calculated by using an average depth multiplied by the area, variations in calculated ocean volume caused by uncertainties in the area can be huge—up to 100 percent for the Arctic in one comparison. The areas used for the world oceans are indicated in Table 2-3.

Individual seas outlined by land masses account for 5.4 percent of the volume and 13.3 percent of the area of the world ocean. No systematic comparison of the chemical behavior of elements within the individual seas with the open ocean is feasible because of wide differences in size, depth, and relationship. Clearly, seas such as the Baltic, Mediterranean, and Gulf of Mexico deserve close examination for the environmental impact of humans—and in some instances they have received such attention. The distinction between surface and deep ocean is also arbitrary and set at a depth of 200 m (10).

The movement of water within the oceans is easy to visualize on a large scale. Circular motions, called gyres, are centered about 30°N and 30°S

TABLE 2-2
Water Reservoirs and Fluxes

	Values (km^3)	Range of Values in Recent Literature (km^3)
Reservoirs:		
Ocean	1,350,000,000	1.32 - 1.37 x 10^9
Atmosphere	13,000	10,500 - 14,000
Land:		
Rivers	1,700	1,020 - 2,120
Freshwater lakes	100,000	30,000 - 150,000
Inland seas, saline	105,000	85,400 - 125,000
Soil moisture	70,000	16,500 - 150,000
Ground water	8,200,000	7 - 330 x 10^6
Ice caps/glaciers	27,500,000	16.5 - 29.2 x 10^6
Biota	1,100	1,000 - 50,000
Flux:		
Evaporation	496,000	446,000 - 577,000
Ocean	425,000	383,000 - 505,000
Land	71,000	63,000 - 73,000
Precipitation	496,000	446,000 - 577,000
Ocean	385,000	320,000 - 458,000
Land	111,000	99,000 - 119,000
Runoff to Oceans	39,700	33,500 - 47,000
Streams	27,000	27,000 - 45,000
Ground feed	12,000	0 - 12,000
Glacial ice	2,500	1,700 - 4,500

Source: Refs. 6, 7, 8, 10, 11, 12, 13, 14, 15.

latitude, with west-flowing equatorial currents that turn poleward at the western boundaries of the oceans. Examples of such currents are the Japan Current in the North Pacific, the Gulf Stream in the North Atlantic, and the Brazil Current in the South Atlantic. Through the mid-latitudes, there is a slow movement of water eastward that is deflected toward the equator near the eastern edges of the ocean. This equatorial flow is associated with upwelling of cold bottom water, resulting in the Canaries Current and the Benguela Current in the Atlantic, and the Peru Current and California Current in the Pacific. The Antarctic Current is not intercepted by land, but continues to flow to the east.

In the Northern Hemisphere, the position of the land masses causes cold

TABLE 2-3
Geographic Distribution of the World Water Cycle

	Area $10^3 \times km^2$	Precipitation km^3	Evaporation km^3	Runoff km^3	Runoff mm	Land with Interior Drainage(%)	Precipitation on Land with Interior Drainage(%)	Continental Runoff km^3	Continental Runoff mm	Ocean Flow Compensation km^3
	A	B	C	D	E	F	G	H		I
Sea	361,110	385,000	424,700	-39,700				-39,700		
North Polar	8,509	826	452	374				2,611		2,985
Atlantic	98,013	74,626	111,085	-36,459				19,351		-17,108
Indian	77,770	81,024	100,508	-19,483				5,601		-13,882
Pacific	176,888	228,523	212,655	15,868				12,137		+28,002
Land	148,904	111,100	71,400	39,700	(266)					
Europe	10,025	6,587	3,761	2,826	(282)	22.3	7.4		(344)	
Asia	44,133	30,724	18,519	12,205	(276)	17.5	12.2		(310)	
Africa	29,785	20,743	17,334	3,409	(114)	28.7	8.7		(397)	
Australia	8,895	7,144	4,750	2,394	(269)	41.0	13.3		(194)	
North America	24,120	15,561	9,721	5,840	(242)	47.2	14.0		(509)	
South America	17,884	27,965	16,926	11,039	(618)	3.7	2.0		(252)	
Antarctica	14,062	2,376	389	1,987	(141)	8.2	2.2		(672)	
							6.0		(141)	
Global	510,014	496,100	496,100							

Source: Data mainly taken from Ref. 8. Runoff (column D) is measured by (Precipitation - Evaporation) (D = B - C) for volume and by (Precipitation - Evaporation)/Area (E = (B - C)/A) for mm values (column E). Columns F and G indicate the amount of runoff that is drained to the interior and column H indicates the intensity of continental runoff in those areas that have runoff to the oceans. The volume to the different oceans is also given in column H. Column I (D + H) indicates the necessary movement of oceanic water to balance the world water cycle.

Arctic waters to move southward through the Bering Strait, Denmark Strait, and Davis Strait, their currents being called, respectively, Kamchatka, Greenland, and Labrador. (See Fig. 2-1.) A profile through the ocean would show surface movement of warm water from the equator toward the poles, and deep movement of cold water from the poles toward the equator. Solubility of CO_2 decreases with increasing temperature (just as warm soda water fizzes giving off bubbles of CO_2), so CO_2 moves from ocean to atmosphere at the equator and the reverse in the polar regions. The concentration of nutrients shows a similar pattern: cold, dense, equator-moving bottom water is high in nutrients, whereas polar-moving water is low in nutrients. This depth-concentration of nutrients is very important for the productivity of organisms, because they use up the nutrients before embarking on the polar trip.

Gyres are centered at those regions of the oceans where evaporation exceeds precipitation and thus causes a net movement of water from the ocean to the atmosphere; this excess of evaporation over precipitation also increases the salinity of the surface water in these areas above the overall ocean average of 35 ppm to almost 36 ppm. The effect of this water loss is initiation of the gyre. Imagine yourself standing on the equator looking east, with the Northern Hemisphere on your left and the Southern Hemisphere on the right. Place your arms in front of you with the palms of your hands facing each other. Curl your fingers inward and raise your thumb. Water is evaporated to the atmosphere—i.e., thumbs point up. The curling fingers toward you are the west-flowing equatorial currents, and the back knuckles are the cold water currents moving from high latitudes toward the equator along the western edges of the continents. These gyres are coupled with atmospheric cells. The evaporated water is distributed by the general wind system and returns to the ocean either by precipitation or through the river system. This is not a "wind driven" system of oceanic circulation but a coupled Coriolis effect caused by the net movement of water out of the ocean into the atmosphere at the major gyres.

Atmospheric Water

The amount of water in the atmosphere is estimated to be 13,000 km^3, an amount that would correspond to a thickness of 25 mm if distributed uniformly on the surface of the Earth. However, the amount of water in the atmosphere is locally variable, ranging from estimates of 1.5 mm over Antarctica to 70 mm in a typhoon over Japan.

The carrying power of wind can move tremendous amounts of material. A 50 km/hr wind moving in a belt 1 km wide and containing 60 mm of water would move 800 metric tons in 1 second. Because moving air masses are hundreds of kilometers across, it is evident that huge amounts of moisture are moved in this manner. For example, 311×10^6 kg/sec of water cross the Pacific coast of North America during the three autumn months and about 250×10^6 kg/sec come across the Gulf coast during the winter months (16). A yearly average of 250×10^6 kg/sec of water moves across the Atlantic coast, with about 110×10^6 kg/sec moving over the Arctic Ocean. Thus, while there is a net excess of precipitation over evaporation, the runoff that results is still significantly less than the amount of water moved through the atmosphere.

A recent Russian calculation (14) indicated that about one-third of the moisture transferred from oceans to land passes over the land and reaches the ocean again. The moisture that does fall is disposed of by evaporation, runoff, or incorporation into other water reservoirs.

The major global wind systems are similar to the ocean currents: Trade winds move from east to west on either side of the equator, separated by doldrums where the air is rising, and mid-latitude westerlies move from west to east between 30° and 70° N and S, with possible high speeds (100 m/sec). Air moving toward the surface beyond 30° moves poleward; particles introduced at 30° or beyond will therefore tend to drift poleward, whereas particles introduced at the equator could drift either north or south.

Residence Time

The preceding discussion of evaporation and precipitation implies a flux, a transfer of material from one reservoir (the atmosphere) to another (the surface ocean). The total amount of precipitation distributed across the whole Earth is estimated to be 2.7 mm/day, which suggests that in less than 10 days the atmosphere has had its water removed, or that water resides in the atmosphere for less than 10 days. Residence time is the expression used to describe the behavior of an element in a particular reservoir. It is defined as

$$\tau = \frac{Q}{dq/dt}$$

where Q is the total amount of element present in the reservoir and dq/dt is the composition of the flux into or out of the reservoir at any specific time. If the influx and outflux are not the same value the residence time of the elements will depend on what value is chosen. If the flux changes, the residence time of all of the elements present in the reservoir changes according to the above definition. It can be seen by this discussion that residence time is a very useful device for a steady-state system: influx = outflux = constant. The problem is that steady-state conditions occur only under strictly defined and temporarily limited natural conditions.

A variety of residence times for water in the atmosphere and ocean can be calculated with differences between output and input. For example, water added to the atmosphere by evaporation from the ocean will have a residence time of $(13,000/425,000) \times 365 = 11$ days, whereas water that is precipitated from the atmosphere has a residence time of only $(13{,}000/496{,}000) \times 365 = 9.5$ days (see Table 2-2). Water in the whole ocean has a residence time of 255 years if it is assumed that all the water evaporated comes from the surface reservoir, the top 500 cm. Residence time varies as the choice of flux varies; when total precipitation (output) and total evaporation (input) are considered, the calculated residence times are approximately the same, 10 days. The short residence time of water in the atmosphere emphasizes and focuses attention on the possible movement of pollutants.

Residence time is sometimes confused with renewal time. Garrels, Mackenzie, and Hunt (17, pages 17–18) use the example of a steady-state bathtub:

> ... 50 gallons of stored water and 5 gallons per minute influx and efflux, the residence time of water of 10 minutes in the tub is obviously the time required to add or subtract an amount of water equal to that in the tub. Does this mean that originally, dirty water in the tub would be completely replaced by clean water from the tap in 10 minutes? The answer could possibly be yes, but in most real situations is no. A condition that would completely renew the water would be one in which the new, clean water did not mix at all with the old, dirty water, but simply displaced it out the drain. One physical situation that would approach this ideal would be to have the tub full of warm, dirty water, and to bring in clean, cold dense water at the bottom, thus pushing the old water out of an overflow at the top of the tub. The reverse of this situation could be imagined in which new clean water disappeared directly down the drain, leaving the amount of dirty water in the tub unchanged.
>
> In many natural reservoirs an intermediate situation occurs, one represented in the bathtub example by continuously and thoroughly mixing the new, clean water with the old, dirty water. At the end of 10 minutes the water in the tub would be a 50-50 mixture of new and old water, so the tub water could be said to be half purified. At the end of a second ten minutes, ¾ of the water would be clean, new water, and so on. For this complete mixing situation, the percentage of original water would diminish logarithmically, with half of the remaining dirty water removed every ten minutes. Just as for the decay of radioactive substance, a *half-life* of ten minutes could be assigned to the original water. At the end of an hour, or 6 half-lives, of this "complete mixing steady state," the fraction of original water would be
>
> $$1/2 \times 1/2 \times 1/2 \times 1/2 \times 1/2 \times 1/2 = 1/64$$
>
> It is common practice to regard 1/64 as an insignificant remaining fraction, and to say that in the complete mixing situation, 6 half-lives or residence times are sufficient to *renew* the reservoir.

In this example the mixing was very rapid, say 10 seconds, and the new and old water were intermingled. If the mixing is slow compared to input and output, then the new and old water have different ratios in different parts of the reservoir. By extension, uniform compositions in the reservoir indicate mixing times much more rapid than residence times.

Land Reservoir

By far the largest surface reservoir of water other than the oceans is that of ice caps and glaciers, with estimates varying from 16,500,000 km^3 to 29,200,000 km^3. More than 80 percent of the ice is in the south polar region and another 10 percent is in Greenland. Part of the variation is due to inaccurate (or lack of) conversion of ice volume to equivalent water volume, and part to increased knowledge of the thickness of the ice packs. This amount of ice is equivalent to the runoff of all the world streams for 900 years. Glaciers can be considered a special kind of river and ice packs a special kind of lake (13). The discharge of water to the oceans is about 2500 km^3 per year (less than 10 percent that of streams) and the residence time for water that is present as ice in Antarctica is approximately 9500 years. Because of minimal mixing of the ice, ice that is millions of years old has been identified and used to show changes in chemistry of the atmosphere and oceans during past glacial ages. Residence times in valley glaciers can vary from tens of years to hundreds of years.

In addition, permanently frozen soil and ground water are present throughout much of North America and Siberia—the permafrost. Permafrost water generally does not enter into the cycle unless it is directly perturbed by humans, and its exact behavior is not yet well known.

Streams, lakes, and inland seas contain approximately 207,000 km^3 of water, or about 16 times as much as the atmosphere. Inland seas and saline lakes account for about half of that amount, with the Caspian Sea making up 75 percent of that half. Evidence shows that different climates in the past have caused major runoff into areas having no ocean drainage, and that the Great Basin of the western United States and similar areas in Australia once had major inland seas. Because water leaves these lakes only through evaporation, they are good objects for studying the volatility of elements.

The bulk of the fresh water is concentrated in a few large lakes. Lake Baikal in Asia, the Great Lakes in North America, and the East African lakes account for almost 90 percent of the fresh lake water. Lakes, along with reservoirs, provide a stabilizing function on runoff. The total amount of water present in streams at one time is estimated to be 1700 km^3 or about 15 percent of the total instantaneous volume of water in the atmosphere. This amount is equivalent to the runoff received by the ocean in about three weeks. The variation of amount present in any particular river at different times is striking, and will be addressed in later sections on rivers as an agent of flux.

Water under the surface of the ground is of three types: soil water, vadose water, and ground water. Soil moisture is of immediate biological concern, as it is the prime source of water for plants—although not all of it is available to the plants. Soil moisture is the buffer between precipitation and ground water. Vadose water lies in the zone of nonsaturation, i.e., not all the pores in the earth material are filled with water. Ground water is defined as water in the zone of saturation, where all pores are saturated. Estimates of the amount of water in each of the three types are only rough calculations. The depth to which soil moisture is affected by precipitation averages about 105 cm, which might be used to define the boundary between soil water and vadose water (18). On the other hand, some plant roots extend 5 m into the ground, indicating that the moisture at that level is eventually transpired to the atmosphere. The amount of soil water is estimated to be about 25,000 km^3 (13)—significantly greater than the 1700 km^3 in streams, and double the 13,000 km^3 in the atmosphere. Estimates of the volume of vadose water are about 45,000 km^3. By far the largest underground reservoir is that of ground water, with estimates ranging from 7,000,000 km^3 (13) to 330,000,000 km^3 (10) including the pore water of sediments. Generally, the value for all three types of underground reservoirs is estimated at 8,100,000 km^3 to 8,400,000 km^3, and given here as 8,270,000 km^3 (Table 2-2).

Discharge of ground water to the sea has been estimated to comprise about 5 percent of the stream flow. Thus, the residence time of water in ground water that discharges to the sea is approximately 4000 years. Most ground water enters stream channels within the continents and accounts for 30 percent of the stream runoff (11). This "nonflood," or stable, stream runoff is raised to 36 percent with the addition of surface water behind dams and lakes. The problem of evaluating ground-water contributions to streams is discussed in a series of chapters of U.S. Geological Survey Professional Paper 813 (for example, 19). The water in streams during periods of high flow is derived from both surface runoff and ground-water discharge. As streamflow decreases, the percentage of streamflow derived from direct surface runoff also decreases. The flow that is equaled or exceeded 90 percent of the time is generally assumed to be totally supplied by ground water. Seasonal variation in the base flow is assumed to be dependent on vapor loss through evapotranspiration. For example, vegetation on flood plains is situated between ground-water recharge areas and stream channels and therefore intercepts the ground water before it is discharged. The cycle of vegetation through the year controls this interception, which is more important in temperate than tropical climates. The 90 percent flow is an indication of base flow in a stream during summer, when vapor discharge is at a maximum, and the 40 percent flow-duration value is an indicator during winter, when evapotranspiration is at a minimum. However, as this has nothing to do with the mean annual flow, it is difficult to construct from past data collections. For example, if the 90 percent flow (discharge of liquid is low) is 650 mgd and the 60 percent flow (vapor discharge is low) is 2460 mgd, then the seasonal variation in ground-water discharge is 1810 mgd (60 percent flow minus 90 percent flow). If the 60 percent flow is a legitimate measure of ground-water discharge, as proposed by the U.S. Geological Survey, it is also a measure of ground-water recharge. In some areas, mining of ground water is now used as a prime source instead of a supplementary source, so that both new and fossil ground water are being added to the water used. Ground-water movement has become critically important in evaluating the movement of pollutants. The chemistry of ground water is highly variable, and is strongly influenced by host rock, especially if it is carbonate (see for example, Table 3-4).

Land Biota

Animal tissues are largely water, and plants contain much water—the estimate used here is 1100 km^3. The importance of water to biota will be discussed later. The importance of biota to water is striking. Transpiration water cannot be distinguished from evaporation water, so there is no way to be certain of the amount of water that cycles through the biota. However, evaporation from land area is 71,400 km^3 and, if we assume that one third moves through vegetation, the residence time is about $(1100/23,800) \times 365 = 17$ days, an amount almost equal to the stable stream mentioned previously.

Fluxes

River

It is important to stress again that the water budget, while balanced over the whole Earth, is highly variable locally. The continents have an excess of precipitation over evaporation of 39,700 km^3/yr (Table 2-3). This runoff balances the deficit of the ocean reservoir, but very unevenly because the Northern Hemisphere is about 50 percent land area and the southern is only 25 percent land area. Each continent is also different. The intensity of runoff can be calculated by dividing runoff by area, giving a depth of runoff per year per km^2 (Table 2-3). This indicates that Africa and Antarctica are very dry (114 mm and 141 mm); South America is very moist (618 mm); and Europe (282 mm), Asia (276 mm), Australia (269 mm), and North America (242 mm) have about the same intensity of overal runoff. However, this is misleading because a significant portion of the land (22 percent) does not drain to the ocean. Land with interior drainage is also distributed unevenly, with North America having about 5 percent, South America about 10 percent, Europe about 20 percent, Asia about 30 percent, Africa about 40 percent, and Australia about 50 percent. In addition, rainfall is distributed unevenly, with less than 8 percent of the total precipitation occurring on land having interior drainage. When the intensity for the ocean runoff areas is calculated, it shows that Australia (509 mm) and Asia

(397 mm) are the continents that seem to vary the most, ranging from areas that receive practically no water to areas of high runoff. The bulk of the European interior runoff goes to the Caspian Sea.

Because runoff to the oceans is a critical parameter of the natural cycles it is important to determine which oceans the runoff feeds and from which continents it comes (Table 2-4). About half of the total runoff goes to the Atlantic (48.7 percent) and about half of that (49.9 percent) comes from South American runoff. It is interesting to note that both the Atlantic and Pacific oceans receive about 70 percent of their inflow on their west coasts, whereas the Indian Ocean receives about 75 percent from its east coast. Table 2-5 lists the major rivers by discharge, and their drainage basins are illustrated in Fig. 2-1. During the year, rivers have fluctuations in discharge, which affect the transport of sediment (see next section). Note that the Amazon River, alone, accounts for 15 percent of the total discharge of water (3767.8 km^3/yr) to the world oceans.

The total land area drained by the major rivers amounts to 42 percent of the continental masses. The intensity of the discharge is calculated by dividing discharge by basin area. For the major river basins it is as follows: Irrawaddy (1029 mm), Ganges (899 mm), Mekong (664 mm), Orinoco (594 mm), Amazon (534 mm), Brahmaputra (509 mm), Fraser (512 mm), Yangtze (353 mm) and Congo (340 mm). The first six are in monsoon areas, showing the inefficiency of water distribution there. The Fraser River is dependent on glacier melting. The remaining columns of Table 2-5 are discussed in the next chapter.

The uneven annual distribution has not been factored into the calculation. If, for example, 50 percent of the annual runoff for the Mekong occurs within two months, during that time the intensity is increased from the annual average of 664 mm to almost 2000 mm. Any understanding of river water as a flux must consider these variations. Unfortunately, adequate data are not available.

River-Ocean Interface

The two types of river-ocean interface are lagoonal and estuarine (17). In lagoonal, fresh-water input is small and sea water enters as a surface flow. In regions where evaporation exceeds precipitation the salinity of the sea water increases (from 35 ppm to 37 ppm in the Mediterranean Sea; Ref. 17). These heavier waters, generally with pollutants dissolved in the Mediterranean water, flow out as an undercurrent at Gibraltar and form a discrete layer in the Atlantic Ocean that has been found thousands of kilometers from Gibraltar.

In the estuarine interface, fresh water flows over the sea water which moves along the bottom as a result of the tides. As stream water mixes with salt water, flocculation of the suspended sediments occurs. The larger particles settle out but are carried landward by the sea water underneath. They accumulate on the floor of the estuary in a fairly narrow zone; that is, there is maximum deposition of suspended sediments from the river in the estuary rather than the sea. These

TABLE 2-4
Distribution of Continental Runoff: Source of Runoff to Oceans

Distribution of Continental Runoff	Amount Runoff km³	Percent to Ocean from Continent				
		North Polar	Atlantic	Pacific	Indian	
Europe	2,564	1.4	98.6			
Asia	12,467	18.1	1.7	51.8	28.3	
Africa	3,409		82.7		17.5	
Australia	2,394			75.7	24.3	
North America	5,840	5.4	61.8	32.8		
South America	11,039		87.5	12.5		
Antarctica	1,987		36.6	28.8	44.6	
Total	39,700	6.6	48.7	30.6	14.1	

Sources of Runoff to Oceans	Amount Runoff Added Km³	Percent from Continent to Ocean					
		Europe	Asia	Africa	Australia	North America	South America	Antarctica
Atlantic	19,351	13.1		14.5		18.7	49.9	2.7
Pacific	12,137		53.2		14.9	15.8	11.4	4.7
North Polar	2,611	1.4	86.6			12.0		
Indian	5,601		63.1	10.7	10.4			15.8
Total	39,700	6.5	31.4	8.6	6.0	14.7	27.8	5.0

Source: Information from Ref. 8.

TABLE 2-5
General Features of the World's Biggest Rivers

River	Length (miles)	Basin area (km³)	Discharge (km³/yr)	Intensity (mm/yr)	Dissolved transport, Td (t/km²/yr)	Solid transport, Ta (t/km²/yr)	Ta/Td	Amount transported (t×10⁶/yr)
A	B	C	D	E	F	G	H	I
1. Amazon	3,915	7,049,980	3,767.8	534	46.4	79.0	1.7	290.0
2. Congo	2,716	3,690,750	1,255.9	340	11.7	13.2	1.1	47.0
3. Yangtze	3,434	1,959,375	690.8	353	NA	490.0	NA	NA
4. Mississippi-Missouri	3,860	3,221,183	556.2	173	40.0	94.0	2.3	131.0
5. Yenisei	3,100	2,597,700	550.8	212	28.0	5.1	.2	73.0
6. Mekong	2,600	810,670	538.3	664	75.0	435.0	5.8	59.0
7. Orinoco	1,283	906,500	538.2	594	52.0	91.0	1.7	50.0
8. Parana	2,406	3,102,820	493.3	159	20.0	40.0	2	56.0
9. Lena	3,636	2,424,017	475.5	196	36.0	6.3	.15	85.0
10. Brahmaputra	1,000	934,990	475.5	509	130.0	1,370.0	10.5	75.0
11. Irrawaddy	1,000	431,000	443.3	1,029	NA	700.0	NA	NA
12. Ganges	2,035	488,992	439.6	899	78.0	537.0	6.9	76.0
13. Mackenzie	3,421	1,766,380	403.7	229	39.0	65.0	1.7	
14. Ob	2,700	3,706,290	395.5	107	20.0	6.3	.3	50.0
15. Amur		1,843,044	349.9	190	10.9	13.6	1.1	20.0

16. St. Lawrence	1,560	1,010,100	322.9	320	51.0	5.0	.1	54.0
17. Indus	1,800	963,480	269.1	279	65.0	500.0	8.0	68.0
18. Zambezi	1,700	1,329,965	269.1	202	11.5	75.0	6.5	15.4
19. Volga	2,291	1,379,952	256.6	185	57.0	19.0	.3	77.0
20. Niger	2,600	1,502,200	224.3	149	9.0	60.0	6.7	10.0
21. Columbia	1,214	668,220	210.8	316	52.0	43.0	.8	34.0
22. Danube	1,777	816,990	197.4	242	75.0	84.0	1.1	60.0
23. Yukon	1,979	865,060	193.8	224	44.0	103.0	2.3	34.8
24. Fraser	850	219,632	112.4	512	NA	NA	NA	NA
25. San Francisco	1,987	652,680	107.7	165	NA	NA	NA	NA
26. Hwang-Ho(Yellow)	2,901	1,258,740	104.1	83	NA	2,150.0	NA	NA
27. Nile	4,157	2,849,000	80.7	28	5.8	37.0	6.4	10.0
28. Nelson	1,600	1,072,260	76.2	71	27.0	NA	NA	31.0
29. Murray-Darling	3,371	1,072,908	12.6	12	8.2	30.0	13.6	2.3

Source: Columns B, C, and D from Ref. 20; Columns F, G, H, and I from Ref. 21. Column E is D/C. NA is not available. See Figure 2-1.

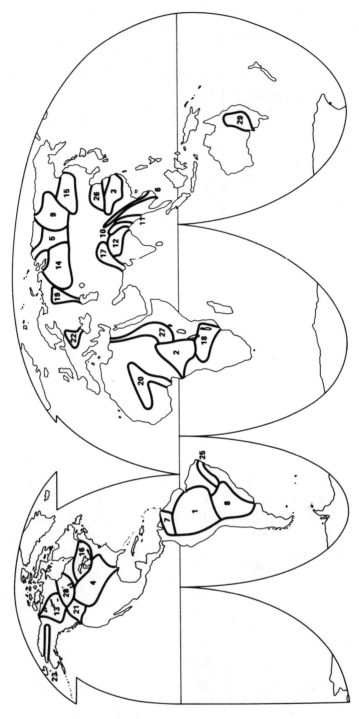

FIGURE 2-1. Major river drainage basins of the world. See Table 2-5 for river identification. Source: Ref. 24, p. 101.

sediments are tremendous sinks for heavy metals and organic compounds (see next chapter). Dredging the sediments to keep shipping lanes clear in the estuary may remobilize the metals attached to the sediments.

Estuarine studies are becoming more common as the realization has grown that the trace elements so mysteriously absent from the ocean and ocean sediments must be present and concentrated in the estuaries (22, 23).

Ocean to Ocean Movement and Residence Time

The movement of water as a function of the evaporation-precipitation differential has already been discussed. We stressed that runoff from the land balanced the excess evaporation, as indicated in Tables 2-2 and 2-3. These tables and Table 2-4 indicate that in the North Polar Ocean there is a surplus of 400 km^3/yr of excess precipitation, which, when coupled with Asian river runoff (75 percent), Mackenzie River runoff (11 percent), and runoff from northern Europe (19 percent), amounts to an addition of 3000 km^3 annually. In the Atlantic Ocean, the water deficit of 36,500 km^3 is not balanced by the river discharge (19,300 km^3), so there is a deficit of 17,200 km^3. This could be balanced by the excess 3000 km^3 from the Arctic and 14,200 km^3 from the Pacific, where there is a surplus. Although actual amounts must be known in order to determine the speed of mixing and the residence times, calculations based on the net effect should give maximum residence times.

Table 2-6 indicates the order of magnitude of residence times of water in the different parts of the world ocean. Line D indicates the net water-movement compensation discussed above, where the minimal flux provides the maximum residence time (time = amount/flux). The uncertainty of exact oceanic volumes does not change the orders of magnitude. Note that the total compensation time for water in the Indian, Atlantic, and Pacific oceans is 20,000-25,000 years; but it is only 3000 years for the North Polar Ocean, thus showing very rapid movement of water through the North Polar seas. The residence time, assuming that all compensation takes place within the surface reservoir of the ocean (upper 200 m), has also been calculated (Line E). The rationale for using this upper zone is that the precipitation-evaporation interchange takes place within it, and the river discharge is added to it. As mentioned previously, because of oceanic mixing these values are higher than the 570-1250 years calculated.

The standard method of calculating residence time is to determine the stream runoff as a measure of flux (Line F). This calculation (Line G) illustrates the effect of the disproportionately large runoff into the Atlantic Ocean discussed previously, and indicates that the introduction of elements by streams would quickly move them through the North Polar and Atlantic oceans relative to the Indian and Pacific oceans. For surface-only calculations, this disparity is even more striking, with both the Atlantic and North Polar showing less than 1000 years (Line H). For water away from land in midocean, it might be reasonable to calculate the residence time by the difference between evaporation and

TABLE 2-6
Oceanic Water Residence Times

	North Polar	Atlantic	Pacific	Indian	Total
A. Ocean volume ($km^3 \times 10^6$)	8.85	350	695	295	1,349
B. Surface ocean volume (200m) ($km^3 \times 10^6$)	1.7	19.6	35.4	15.5	72.2
C. Ocean flow compensation (km^3/yr)	3,000	-17,100	+28,000	-13,900	0
D. Total compensation time: A/C (yr)	2,950	20,500	25,000	21,200	
E. Surface only compensation time: B/C (yr)	570	1,150	1,250	1,100	
F. Stream runoff to oceans (km^3/yr)	2,600	19,400	21,100	5,600	39,700
G. Runoff residence time: A/F (yr)	3,400	18,000	57,500	52,700	34,000
H. Surface runoff residence time: B/F (yr)	650	1,000	2,900	2,800	1,800
I. Atmospheric cycling (precipitation minus evaporation) (km^3/yr)	400	-36,500	15,900	-19,500	-39,700
J. Whole ocean atmospheric cycling residence time: A/I (yr)	22,125	9,600	43,700	15,100	34,000
K. Atmospheric cycling surface ocean residence time: B/I (yr)	4,250	500	2,200	1,000	1,800

Source: Ref. 24.

precipitation. The Atlantic loses significantly more water through evaporation than it gains through precipitation. Thus, nonvolatile elements would be expected to be concentrated in the ocean, and volatile elements quickly released to the atmosphere. This release of volatiles would be expected to be insignificant in the North Polar relative to water flow, it would be minor in the Pacific, but it would be of major importance in the Indian and Atlantic oceans. It is important to emphasize the predicted net movement of water from the Pacific and North Polar to the Atlantic and Indian oceans. However, this movement probably does not have a great diluting effect on the bordering seas, bays and estuaries that appear to be the problem areas of pollution (22).

Atmosphere-Land Interface

The interchange of water between soil biota and atmosphere starts with precipitation (18, 19). The part that wets the above-ground vegetation and evaporates when the rain ceases is termed intercepted rain (I); it is not available for transpiration by plants or for ground-water recharge. The greater the canopy cover, the greater the amount of interception; the amount varies with tree species (being greater for spruce than for beech) and with the amount of rainfall until capacity is reached. For spruce, 6 mm of rainfall is necessary to fill its 3-mm interception capacity. The rain that is not intercepted reaches the ground in three ways: throughfall (Pt), canopy drop (Pd), or stem flow (Ps). Only the throughfall has the same properties as the rain above the canopy, and it can be distinguished from canopy drip only in the open. Thus, the total amount of precipitation can be expressed as:

$$P = I + Pt + Pd + Ps$$

Stem flow can be extremely effective, and a rainstorm may produce 500 l of water flowing down the stem and entering the soil at the base of a dominant beech tree whose branches and stems rise and have smooth surfaces—which facilitates such flow. Spruce, on the other hand, has falling branches that lead rain water to the periphery, where it drops off, as shown below (18).

Spruce	Percent
I	25
$Pt + Pd$	75
Ps	
Beech	
I	9
$Pt + Pd$	76
Ps	14

Finally, if the precipitation falls as snow, the effects are different. Interception is large but evaporation is low, and most of the snow either plops or drips to the ground, thus decreasing the true interception.

After reaching the ground, precipitation tends to infiltrate the soil as long as the input rate does not exceed the "infiltrability"; if it does, surface runoff is possible. The organic, upper soil layer in the forest provides a high storage capacity, and this minimizes runoff. The infiltrating water increases the amount of water in the soil pores, and provides the reservoir from which plant roots extract water. The fate of the entering water depends on the soil's water-retention ability and water conductivity. The total pore space of a soil consists of pores of various sizes, and the strength of the forces holding the water in the soil decreases with increasing pore size. If air and water are both present, air occupies the larger pores and the system is called unsaturated. At least 10 percent of the total soil volume should be filled by air in the main root zone to provide sufficient air for the plant. (A saturated soil has insufficient air for plant growth.)

Because most of the pore space is of capillary dimension, capillary force causes a soil-water potential, called soil-water tension, capillary potential, or matrix potential. This potential is the force that causes the water to move. Such water movement is in the direction of negative hydraulic gradient, and is proportional to such gradient by a coefficient called permeability or soil-water conductivity. The coefficient is highly dependent on soil structure and the degree of water saturation of the soil.

Subsurface flow can occur in any direction between horizontal and vertical. The vertical aspect of such flow, termed deep seepage, is water that moves out of the root zone, is a primary soil leaching agent, and replenishes the ground water. It is interesting to note that although the surface and subsurface flows have been considered as outputs, they can easily be inputs, either artificially (through irrigation) or naturally (downhill portion of slope).

The remaining outputs are evaporation and transpiration. Evaporation removes the water from the surface (using about 590 cal/cm^3 water), creating a hydraulic gradient that moves the moisture upward through the soil profile to the surface. Evaporation is a minor component under forest cover, partly because the low permeability of the organic layer limits upward movement despite steep vertical hydraulic gradients.

Transpiration can best be considered as a complete process that starts in the soil as soil-water potential and ends in the atmosphere as vapor pressure. Such potential deficiencies can amount to as much as hundreds of bars during sunlight hours. As long as the evaporative demand of the atmosphere can be met, the role of the plant is passive. It absorbs water in the root zone at a rate dominated by the energy available at transpiration surfaces, rather than by the plant's physiological requirements. When soil moisture is limited, plants begin to regulate the transpiration flux, generally wilting irreversibly when the

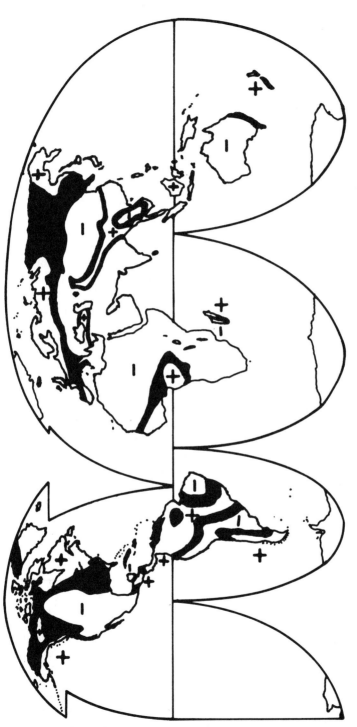

FIGURE 2-2. Regions of surplus and deficit of river water resources (± 200 mm). Source: Modified from Ref. 15, Map 6.

soil-water potential is less than −15 bars. The transpiration rate is a function not only of the matrix potential of the root zone but also of the evaporation demand caused by weather conditions. Water that still exists in the soil at less than −15 bars soil-water potential—the level that supports plant transpiration—is termed unavailable (dead) water.

The atmosphere-land interface controls the water balance dynamics through the interactions of climate, soil, and vegetation discussed above. A statistical dynamic formulation of water balance in terms of annual precipitation, potential evapotranspiration, and physical parameters of soil, vegetation, climate, and water table has recently been developed (25). Several tests indicate the applicability of the model for particular regions.

There are several generalized correlations of precipitation, evaporation, vegetation type, latitude, and runoff that guide predictions for particular basins (7, 8, 14). Maps of these variations have been prepared for UNESCO (15). The difference between precipitation and evapotranspiration is a measure of the amount of possible water runoff, and thus is a measure of world regions with a surplus or deficit of river water resources. These areas are indicated in Figure 2-2. Regions with a significant surplus (annual precipitation is greater than annual evapotranspiration by more than 200 mm; indicated by +) are concentrated in southeast Asia, the Amazon and Orinoco basins, northeastern United States and Canada, northern Europe, northeast Asia, and the Congo basin in Africa. Water deficit regions (precipitation is less than evapotranspiration by at least 200 mm; indicated by −) are shown to dominate Australia, southern Asia, nearly all of Africa, the western half of North America, and large areas of South America. The shaded regions indicate areas of generalized water balance that could be sensitive to short-term climate effects. The patterns of movement of geochemical material are tremendously influenced by the surface availability of runoff water. These patterns are discussed in the next chapter.

References Cited

1. Murray, C. R. and E. B. Reeves, 1972, Estimated Water Use in the United States in 1970; U.S. Geological Survey Circular 676, Washington, D.C., 37 p.

2. Murray, C. R. and E. B. Reeves, 1977, Estimated Water Use in the United States in 1975; U.S. Geological Survey Circular 765, Washington, D.C., 39 p.

3. Leifeste, D. K., 1974, Dissolved Solids Discharge to the Oceans from the Conterminous United States; U.S. Geological Survey Circular 685, Washington, D.C., 8 p.

4. Vander Leeden, F., 1975, Water Resources of the World; Water Information Center, Port Washington, N.Y., 568 p.

5. National Water Commission, 1973, New Directions in U.S. Water Policy; Summary, Conclusions, and Recommendations from the Final Report of the National Water Commission; National Water Commission, Washington, D.C., 197 p.

6. Lvovitch, M. I., 1977, World Water Resources Present and Future; AMBIO, v. 6, no. 1, p. 13-21.
7. Lvovitch, M. I., 1979, World Water Resources and Their Future; English trans., R. L. Nace, ed., American Geophysical Union, Washington, D.C., 415 p.
8. Baumgartner, A. and E. Reichel, 1975, The World Water Balance, Mean Annual Global, Continental, and Maritime Precipitation, Evaporation, and Runoff; trans. by Richard Lee, Elsevier, New York, 179 p.
9. Menard, H. W. and S. M. Smith, 1966, Hypsometry of Ocean Basin Provinces; Journal of Geophysical Research, v. 71, p. 4305-4325.
10. Garrels, R. M. and F. T. Mackenzie, 1971, Evolution of Sedimentary Rocks; W. W. Norton & Co., New York, 397 p.
11. Lvovitch, M. I., 1973, The Global Water Balance; EOS, v. 54, no. 1, p. 28-42.
12. Nace, R. L., 1967, Water Resources: A Global Problem with Local Roots; Environmental Science and Technology, v. 1, no. 7, p. 550-560.
13. Nace, R. L., 1969, World Water Inventory and Control; p. 31-42 *in* Water, Earth, and Man, R. J. Chorley, ed., Methuen, London, 588 p.
14. USSR Committee for the International Hydrological Decade, U. I. Korzoun, editor-in-chief, 1978, World Water Balance and Water Resources of the Earth; UNESCO, Paris, 663 p.
15. USSR Committee for the International Hydrological Decade, U. I. Korzoun, editor-in-chief, 1978, Atlas of World Water Balance; UNESCO, Paris, 36 p. plus 63 maps.
16. Barry, R. G., 1969, The World Hydrological Cycle; p. 11-29 *in* Water, Earth, and Man, R. J. Chorley, ed., Methuen, London, 588 p.
17. Garrels, R. M., F. T. Mackenzie, and C. Hunt, 1975, Chemical Cycles and the Global Environment; Wm. Kauffman, Los Altos, Calif., 205 p.
18. Benecke, P., 1976, Soil Water Relations and Water Exchange of Forest Ecosystems; p. 101-131 *in* Water and Plant Life, O. L. Lange, L. Kapper, and E. D. Schulze, eds., Springer-Verlag, Berlin, 536 p.
19. Bloyd, R. M., Jr., 1974, Summary Appraisals of the Nation's Ground-Water Resources—Ohio Region; U.S. Geological Survey Professional Paper 813-A, Washington, D.C., 41 p.
20. Browzin, B. S., 1972, Rivers; p. 353-362 *in* Encyclopedia Britannica, v. 19, Britannica Corp.
21. Meybeck, M., 1976, Total Mineral Dissolved Transport by World Major Rivers; Hydrological Science Bulletin, v. 21, no. 6, p. 265-284.
22. Turekian, K. T., 1977, The Fate of Metals in the Oceans; Geochimica et Cosmochimica Acta, v. 41, p. 1139-1144.
23. Förstner, U. and G.T.W. Wittmann, 1979, Metal Pollution in the Aquatic Environment; Springer-Verlag, Berlin, 486 p.
24. Speidel, D. H. and A. F. Agnew, 1979, The Natural Geochemistry of Our Environment; p. 77-239 *in* An Overview of Research in Biogeochemistry and Environmental Health, Committee Print 825, Committee on Science and Technology, U.S. House of Representatives, Washington, D.C.
25. Eagleson, P. S., 1978, Climate, Soil, and Vegetation; Water Resources Research, v. 14, p. 705-776.

3
Geochemical Movement

This world along its path advances.
—Thomas More

Introduction

The movement of elements takes place physically, as mineral grains and/or rock fragments, or chemically, in solutions. Generally the same elements occur in both types of movement in varying proportions. Some of the well-known dispersion mechanisms are given in Table 3-1. Rapid mass movement of material such as occurs in mud slides or rock slides, while locally important, does not contribute a major portion of total geochemical movement. Slow mass movement, such as soil creep in response to gravitational forces or seasonal fluctuations, transports tremendous amounts of material, but for only short distances. It is the transportation of fluids such as streams, wind, and ice that provides most of the movement of interest to the environmental geochemist. Some of this transport, such as capillary movement of soil water and stem flow, discussed in the previous chapter (1), moves elements that are necessary for biological processes. Unfortunately, however, only limited estimates of those quantities are available. For example, the total tree-stem load can exceed the total precipitation runoff from surface water and ground water combined (2). In this chapter we shall concentrate on transportation of elements by streams, ice, and wind.

Transport by Streams

Dissolved Load

Evaluation of the dissolved load can be made by multiplying the stream discharge (39,700 km^3/yr) by the dissolved concentration (or chemical composition) of the stream. The total amount is thus a combined function of the concentration of dissolved ions (salinity) and the amount of discharge. For example, a river with a salinity of 5 ppm and a discharge of 40 km^3/yr will

TABLE 3-1
Geochemical Movement

Type and Agent	Annual Amount	Comments
A. Particle movement		
1. Moving Water	25×10^9 t/yr	See Transport by Streams. Streams are predominate but wave action on coast can be significant.
2. Glacial Ice	$0.2-0.5 \times 10^9$ t/yr	See Ice Transport. Now most important in Antarctica. Icebergs give extreme lateral movement.
3. Wind	$1.6-3.8 \times 10^9$ t/yr	See Transport by the Atmosphere. Loess deposits, desertification effects can be exaggerated.
4. Mass Movement	?	Landslides, rockslides, and soil creep are difficult to quantify for annual movement.
B. Solution Dispersion		
1. Surface Water	5.2×10^9 t/yr	See Transport by Streams. Depends on Eh, pH, nature of acids, solubilities, and mineral sources.
2. Ground Water	?	See Transport by Streams. Movement into streams is included with surface water.
3. Soil Water	?	Capillary movement through soil and plants, return with litter.
4. Stem Flow	?	Qualitatively large but few analyses.

Source: See text sections referred to in Comments.

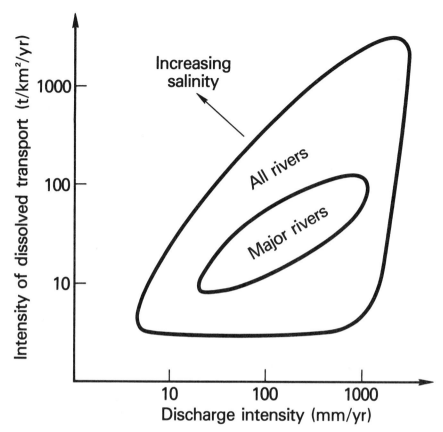

FIGURE 3-1. Relationships of variations in dissolved load of streams with intensity of stream discharge. Source: Ref. 4. See also Table 2-5.

transfer twice the amount of dissolved material as a stream with a salinity of 20 ppm but a discharge of only 5 km³/yr. Figure 3-1 and Table 2-5 indicate that there is an increase in the amount of material moved with increasing river discharge, in spite of the general decrease in salinity with increased discharge. The salinity may be lower, but more material is moved. This inverse relationship of salinity and discharge is a complex function. For example, in some instances floodwaters do not reflect the expected dilution but are characterized by increases in the concentrations of SO_4, Mg, Ca, and K (3).

There is also a relationship between river-basin relief and dissolved transport. Maximum values of river dissolved transport (t/km²/yr) are always in mountain regions (5). The lowest values are in areas of low runoff (the Rio Grande), of high runoff in areas of low relief (the Congo River), or in areas of

TABLE 3-2
Dissolved Load Discharged to World Ocean

Author (Ref.)	Year	Salinity (ppm)	Total Discharge 10^6 t/yr
Clark (10)	1924		3700
Livingstone (9)	1963	120	3760
Alekin & Brazhnikover (8)	1968	88	3600
Garrels & Mackenzie (11)	1971	130	4200
Meybeck (4)	1976	89	3250
Meybeck (5)	1977	108	4103
Holland (7)	1978	117	5380
L'vovich (12)	1979	63	2480
This work	--	130	5160

both low relief and low runoff (the Nelson and Murray rivers). Indeed, river-basin relief is the most important parameter controlling the geochemistry of the Amazon River (6). Thus, the dissolved-ion content cannot be treated as a simple function of discharge intensity, as one author has recently done (7, his Fig. 4-6).

There are few major published reports on worldwide chemical compositions of streams. Dissolved-load data for the Soviet Union, as a function of climate and vegetation type, were extended worldwide by using geographic distribution of vegetation type as a key, arriving at a world value of 88 ppm (8). Another author (9) used published analyses weighted by the area drained by the stream, and arrived at 120 ppm dissolved load (or 130 ppm, including dissolved organics). These numbers are questionable, both because they are based primarily on temperate-zone runoff, and because they do not consider monthly variation. Table 3-2 gives a comparison for worldwide estimates, including more recent compilations. Variations between particular regions can be striking; values of salinity for different rivers can range from about 5 ppm to more than 20,000 ppm. Even continental averages vary greatly, with estimates of 152 ppm for Australia, 55 ppm for South America, and only 28 ppm for the high-discharge area of the Malay Archipelago (6). This variation again emphasizes that any geochemical cycle must be based on the particular region rather than on worldwide values.

TABLE 3-3
Composition of River Water, Worldwide Average for Major Ions in Solution

Ions	Range of Values* (ppm)	Dissolved Transport / Total Transport	Comments
HCO_3^-	57.7 - 58.4	90% (for C)	31 for S. America 95 for Europe
SO_4^{2-}	8.5 - 11.2	85% (for S)	4.8 for S. America 24 for Europe
Cl^-	5.3 - 7.8	75%	4.9 for S. America 12.1 for Africa
Ca^{2+}	14.6 - 15	60%	3.9 for Australia 31.1 for Europe
Na^+	5.1 - 6.3	60%	2.9 for Australia 11 for Africa
Mg^{2+}	3.8 - 4.1	50%	1.5 for S. America 5.6 for Asia

* Values taken from tabulations in References 8, 9, and 13.

Different authors show different composition of dissolved ions and different salinity, varying with area (Table 3-3), but they generally agree as to the major constituents and their relative proportions. For example, another Soviet group (14) estimates that a total of 3.430×10^9 tonnes of dissolved ionic material is transported by streams. Of this amount, 0.273×10^9 tonnes are transported into interior basins. In addition to this amount of ionic material, there are transported 0.218×10^9 tonnes of dissolved or colloidal Si, Al, and Fe; 174×10^9 tonnes of complexed organic carbon; 12×10^9 tonnes of nitrogen and phosphorus; and slightly less than 3×10^9 tonnes of other dissolved elements.

The chemistry of river water can be described in three ways (15): by rock control, by precipitation control, or by evaporation and crystallization control. All are a result of the interaction of topographic relief and climate. In the rock-dominated areas the waters are in partial equilibrium with the material in their basins, and have high $(Ca-HCO_3)/(Na-Cl)$ ratios, as shown in Figure 3-2. With increased rainfall, the dilution caused by increased runoff decreases the salinity. Also, with increased rainfall, the elements added thereby begin to dominate; that is, the proportions of Na and Cl increase relative to those of Ca and HCO_3. More than 80 percent of the dissolved salts in some Amazon

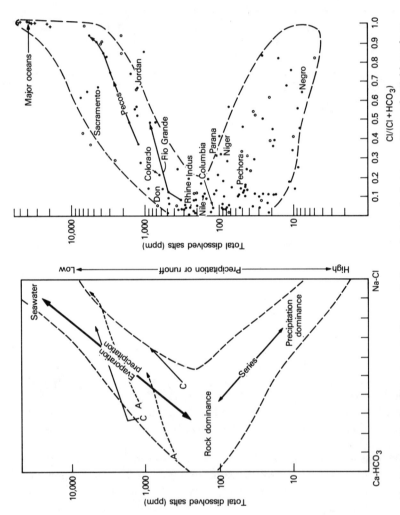

FIGURE 3-2. Mechanisms controlling salinity and chemical composition of world surface water; variation of salinity and Cl/(Cl + HCO$_3$). Source: Ref. 15. Copyright 1970 by the American Association for the Advancement of Science. Reprinted with permission.

tributaries are added through precipitation (15); thus, volatile elements might be concentrated in these rivers and lakes. The other major mechanisms of concentration of Na and Cl are evaporation, which increases salinity by removing water, and crystallization (precipitation) of $CaCO_3$ from solution, which increases the Na-Cl proportions. For rivers of the last type, such as the Rio Grande, more than 99 percent of the ions are contributed from the source rocks. It is interesting to note that high Na-Cl proportions occur in warm to hot climates, high salinities in dry climates, and low salinities in humid climates.

Even in rock-dominated regions there is argument as to whether the stream chemistry is controlled by the conversion of primary materials to secondary kaolinite and montmorillonite, or to other reactions with the rock. For example, one author argues that very large volumes of rock are slightly altered by depletion in ferromagnesian minerals without evidence of clay formation, with the high rate of erosion enabling the alteration process to continue (16).

On a different side of the problem, the presence of Na and Cl in river water is used as an indication of the contribution of ocean spray to the rain water that provided the runoff. But recently (17), the discovery of rock salt deposits in the Peruvian Andes cut by tributaries of the Amazon has been used to account for approximately half of the sodium chloride in the rivers there — a case of rock-dominance where none was expected to exist for chlorine.

The composition of stream water is a complex mixture of the chemistry of rain modified during the runoff process and the chemistry of ground water. Table 3-4 illustrates some of the chemical differences among rain water, stream water, ground water associated with different rock types, and ocean water. Analyses of subsurface water by rock type have shown that the three most important sources of ground water are sandstones, carbonate rocks, and unconsolidated sands and gravels (19). Ground water from alluvium of predominantly igneous origin is similar to water from igneous rocks, but most closely related to surface waters in the same drainage basin. A compositional plot of $(Cl + Na)/(Cl + HCO_3 + Na + Ca)$ falls into the rock-dominance regions discussed for surface waters (Figure 3-2), a result to be expected if the chemistry of the water results from slow reaction with the host material. It is also important to realize that ground water has a pH of about 7, but the pH of precipitation-dominated runoff is less than 6. Solubility relationships change at the interface where ground water enters stream runoff. Ground water is highly variable locally for minor elements and total dissolved elements — for example, the median concentration of strontium in public water supplies is 110 ppb, but for Denver it is 4.1 ppb and for San Diego it is 1100 ppb (20).

The composition of stream water at the 40 percent mean-discharge level can be assumed to be a reasonably accurate representation of the ground-water

TABLE 3-4
Comparison of Water Chemistry

Type of Water	Rain*		River	Ground Water**			Ocean
	A	B		A	B	C	
Ca/Na (weight)	0.34	8.3	2.2	10	4.9	5.2	0.038
HCO_3/Cl (weight)	1.3	16.2	7.0	48	32	24	0.0074
Salinity (ppm)	12.5	3.7	130	420	430	470	34,500
Reference	18	16	8,9	19	19	19	19

* Rain: A = analysis from California
B = analysis from Wyoming

** Ground Water: A = carbonate host rock
B = sandstone host rock
C = unconsolidated sands and gravels

contribution to stream chemistry. Unfortunately, even detailed studies of basins such as the Hubbard Brook region do not tabulate such measurements (21). The result is that we are still unable to predict accurately the stream chemistry of a particular watershed, based on the proportion of contributions from rain runoff and ground water.

Suspended Load

The total suspended load of streams is even more difficult to estimate than the dissolved load, and perhaps more critical. Sediment is an excellent example of the definition of a pollutant — a resource out of place. It depletes the land from which it is derived, it diminishes the quality of the water in which it is transported, and it often degrades the location where it is deposited. Sediment consists of soil and rock particles eroded from both disturbed and undisturbed lands; crop, range, and forest areas; highway rights-of-way; surface-mined areas; stream banks; and construction sites. Sediment fills reservoirs, lakes, and ponds; clogs streams; covers as well as creates productive land; interferes with aquatic habitat; degrades water for other uses; and provides beaches for recreational enjoyment. Loadings of solids reaching streams from surface runoff are estimated to be at least 700 times the loadings from sewage discharge. Runoff is therefore responsible not only for the increased dissolved load, but also for the suspended load.

Estimates of the amount of sediment reaching the ocean annually range from as much as 51.1×10^9 metric tons (22) to as little as 6×10^9 metric tons

(23). The lower value is assumed to be free from human-caused additions, and compares with a dissolved load of 4–5 × 10^9 metric tons. The most commonly used value is 18.3 × 10^9 metric tons of sediment, of which 80 percent comes from Asian river runoff (24). Unfortunately, the sampled area includes less than 40 percent of the land surface of the world. The more recent estimate of 15.7 × 10^9 t/yr by the Soviet group (14) does not seem to consider adequately the large impact of tropical rivers. Another effort, which attempts to derive a sediment-delivery regression equation calculated from topographic and climatological data on known rivers, extended to worldwide coverage (25), gives a value of 26.7 × 10^9 t/yr.

Erosion rates have been shown to be lowest in tropical and continental forest climates, intermediate in arid and semiarid climates, and very high in Mediterranean climates—with the highest values in areas of high altitude and relief. Classic sediment-yield curves (26) are misleading oversimplifications, as has been shown for seasonal variation in a continental climate (27). Plots of sediment yield versus discharge for individual streams generally show a log-linear relationship, but the results for one basin cannot be transferred to another. Small basins have higher sediment delivery than larger ones (see below); mountain streams have higher sediment delivery ratios than do streams in areas of low relief. Seasonal variations in sediment movement are greater in areas of high relief than in those of low relief (6). Generalized patterns can be seen in Table 2-5, however.

The relationship of dissolved transport per unit area (Td) and solid transport per unit area (Ts) of the major rivers of the world, and the variation of Ts/Td by climate and relief are illustrated in Figure 3.3. Figure 3-3 also shows the topographic-precipitation control on dissolved-transport intensity and solid-transport intensity. Area 1 is mountainous (or the mouth of the stream is close to the mountainous source) with high precipitation, examples being the Mekong, Ganges, and Brahmaputra. Area 2 is mountainous with much lower precipitation and longer dry periods, such as the Zambezi. Area 3 has average relief and can be either tropical or temperate, e.g., the Mississippi and Orinoco. Area 4 is dry with low relief, such as the Nile and the Murray-Darling. Areas 5, 6, and 7 all have low relief with temperate, subarctic, and tropical climates, respectively. An example of area 5 would be the Volga; of 6 would be the Lena, the Mackenzie, or the St. Lawrence; and of 7, the Congo.

Seasonal variations within these seven areas are generally similar (28). For example, areas 3 (when tropical) and 7 have late-summer peaks corresponding to the monsoon runoff. The runoff is even more concentrated in the dry river areas 4 or 2, with massive amounts of solids carried in the sudden rush of water. For temperate-climate rivers, the water is more evenly distributed, and Ts/Td is closer to 1. Subarctic rivers have a major flow in May or June, caused by the melting of ice and snow. The magnitude of the variation for a

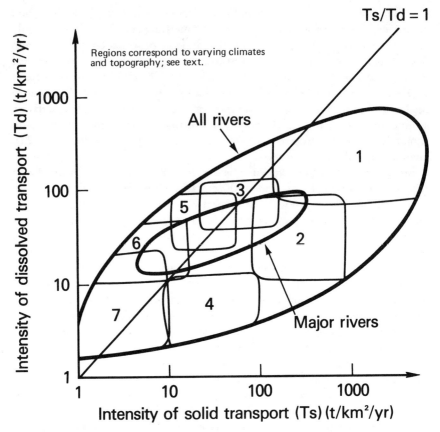

FIGURE 3-3. Relationship of variations in dissolved load of streams with solid load of streams. Source: Ref. 4. See also Table 2-5.

particular river is striking, as are the differences between river regimes. This emphasizes the importance of such work as that of Gibbs (29), who found generally similar behavior for some elements in the load of the Amazon and the Yukon.

The decrease in sediment discharge due to human causes can be striking. For example, an estimate for the silt load of the Nile River in 1950 was 110.5 million t/yr, but for 1973 was only 103.2 thousand t/yr—a huge decrease (a factor of 10^3)—a result of silting behind the Aswan dam in its reservoir. A reservoir having a capacity of only 1 percent of the annual discharge can reduce the sediment load by half (30). Before dam construction, 50-65 percent of the material carried by the Huang Ho in China was redeposited. The most striking example is the load carried to tidewater by the Colorado River in

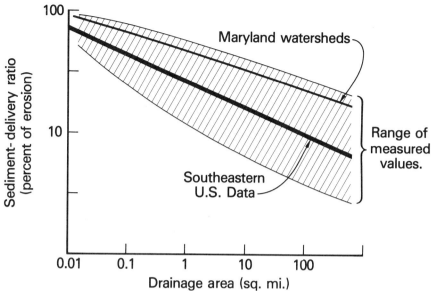

FIGURE 3-4. Effect of size of drainage area on sediment-delivery ratio.
Source: Ref. 31.

the southwestern United States; today it carries no sediment at all because the water is totally consumed by humans before it reaches the river's mouth. The sediments delivered to the ocean are only a small fraction of the total sediment moved by the river each year. It is estimated that only 5 percent of the sediment eroded into the Potomac River eventually reaches tidewater (24).

The sediment-delivery ratio is also clearly a function of basin size (31, 32, 33, 34). A smaller basin has a higher delivery ratio than a larger one — that is, less sediment is deposited within the basin where it is being eroded (see Fig. 3-4). This means the channeling streams have a ready supply of sediment for transport (35).

The distribution and intensity of rainfall have been used to predict the rate of denudation (22, 36). The factor used is p^2/P^2 where p is the maximum mean monthly precipiation and P is the mean annual precipitation. Thus, 120 mm of rain per year, all in one month, would give a factor of 144, whereas if the rain were distributed evenly throughout the year the factor would be one. The factor changes seasonally and increases with total annual precipitation. As mentioned previously, large storms can have a disproportionate effect, as indicated by the Shenandoah River in the southeastern United States (37). Nearly 100 times the amount transported in 1956 was moved in one month in 1955, as a result of a hurricane. It is interesting to note the scouring effects of large storms: for example, in April 1974 the Shenandoah had a much lower sediment discharge than would be predicted from its river flow, but in March 1974 the river had delivered about 80 percent of its annual sediment load. This substantiates the previous statement that sediment deposited in upper reaches is remobilized in times of extraordinarily large runoff. In most temperate regions the episodic event appears to cause most of the material to be transported, when considered over a long time span (29).

Factors other than floods can be important. Reservoir construction and improved land-use practices have not reduced the amount of sediment transported to the Atlantic Ocean by U.S. rivers (38). Tributaries below the dams and the increased carrying capacity of the main stream below the dam are apparently now delivering the sediments that previously had accumulated. This is similar to the problem of the accumulation in San Francisco Bay of sediments originally put into the Sacramento River basin by hydraulic gold mining during the nineteenth century.

Human-caused sources of sediment reaching streams are agricultural tillage, domestic animal grazing, highway construction and maintenance, timbering, mining, urbanization, and recreational land development. The first two are probably the most important quantitatively worldwide (37).

One more factor has to be considered. The behavior of the Susquehanna River in Pennsylvania during Hurricane Agnes in 1972 (39) demonstrates that rivers can drastically change their climate-behavior pattern during extraordinary changes in river flow—i.e., the sediment load carried in extreme floods can be much greater than that predicted on the basis of increased water volume alone. Fifty years of normal sediment load (50,000,000 tonnes) of the Susquehanna River was equalled in a two-week period in 1972 (39). A flood can thus apparently balance the erosion or deposition shortage indicated by the sediment-delivery ratio by quickly removing the sediment that has been deposited in smaller basins (40). A reasonable measure of the potential sediment-delivery ratio can be made by assuming that the ratio of the smaller basins is equal to that of the larger ones. The differences between this value and the normal one would be the sediment delivered by the floodwaters of the extraordinary event. This conclusion is quite different than conventional ideas (34, 41) that 80-95 percent of sediment is moved in events that are more frequent than once a year. Clearly, the catastrophic event moves most of the sediment. This, of course, makes it even more difficult to measure and predict sediment movement.

Can we predict erosion? The universal soil loss equation (42) considers the effects of rainfall as erosive energy, soil erodibility, length and steepness of slope, plant and litter cover, and control practices. The equation curves apply only in the eastern part of the United States, but do have good local applicability for normal runoff and erosion. Since the computations are based on an observed 22-year range, they suffer statistically from the problem of small numbers. Numbers derived this way, coupled with the sediment-delivery ratio of a particular watershed (see Fig. 3-4), give an indication of the amount of sediment "in storage" along the path followed by water runoff.

One more problem is to extend this type of analysis into different climates. For example, vegetation cover in arid climates activates local turbulent flow and induces slope erosion to a degree greater than that observed where there is no vegetation cover (43). Clearly, we are still a long way from predictive capability for sediment movement.

It is difficult to relate the behavior of a particular element to its distribution between suspended and dissolved load. The mobility of particular elements is discussed in the following chapter.

Estuaries: Interface of Rivers with the Ocean

When a large amount of river water enters an estuary, the latter acts mainly as a conduit; when the discharge is small, the estuary acts as a special standing body with its own properties and behavior. The cross-sectional area used to measure the discharge at the interface of river water and sea water is commonly taken as the contour line of 1 ppt (1000 ppm) salinity (the brackish-water line). The flushing velocity is obtained by dividing the discharge by the area. This flushing velocity is a measure of the river's ability to keep sea water out of the estuary, and is also a measure of the ability of the river to deliver its load through the estuary to the ocean. Estuaries tend to retain the fine sediment, thus providing a reservoir for heavy-metal pollutants and serving as an inhibitor for benthic life forms. For example, the Susquehanna River contributes 1,070,000 tonnes of sediment per year to the Chesapeake Bay, an additional 600,000 tonnes come to the bay from shore erosion, and 220,000 tonnes more are brought into the bay from the ocean—which is at least an indication of the low flushing velocities of the rivers into the bay (44).

By assuming a constant density, a volume of 860,000 m^3/yr can be calculated as the amount of sediment that moves through the Chesapeake Bay estuary. This number is so huge that it creates a conceptual problem in grasping the effects of erosion. An alternate way of expressing this amount is to determine the thickness that the sediment would reach if it were spread over the whole bay (which it obviously is not), resulting in a baywide value of 0.8 mm/yr or 80 $cm/10^3 yr$. Thus we can move conceptually from very large numbers of overwhelming size to ones that appear insignificant.

The change of sediment concentration suspended in water as one moves from mountain rivers in flood to water over the continental shelf beneath the ocean is indicated in Figure 3-5. The range of sediments is seen to be over 7 orders of magnitude. The source of the bottom sediment on the inner shelf is not yet agreed upon, but it could be delivered by floods, as shown by previous discussion. Other sources could be resuspended bottom sediments or organic

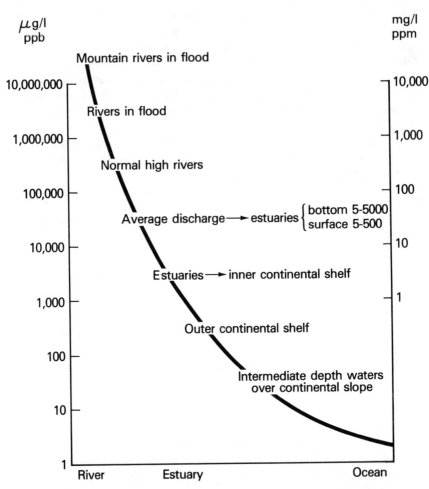

FIGURE 3-5. Change in concentration of suspended matter in different waters. Sources: Refs. 30, 45, 46.

material from the ocean surface. Some samples taken in the lower Susquehanna during low and moderate flows are not representative of the bulk of the material discharged (47). It is likely that samples taken at the estuary-ocean interface would also not be representative of the bulk of the material discharged to the shelf.

Diagrammatic representation of the natural cycle of sediments in the coastal zone (48) indicates that nonriver sources are not likely to add sediment that possesses initial concentrations much greater than the natural cycle. It is also known that the amount moved in storms appears to be mostly independent of man's actions. The suspended sediment generally shows a concentration maximum on the landward side of the salt-water boundary (1000 ppm). The sediment consists not only of particles settling out of the upper layer and carried landward, but also finer particles flocculated by the action of the sea. If the flushing velocity is low, the turbidity maximum will be in the upper part of the estuary and never be exposed to the oceanic transporting agents (49). As the magnitude of the tidal flow increases relative to the river flow and/or the width of the basin increases relative to its depth, the estuary changes character from stratified to mixed, thus reducing the estuary's sediment-trapping capability. An estuary's sediment-trapping efficiency is greatest when river flow is highest, and therefore when fluvial sediment input is normally greatest. It is also significant that dredging increases the trapping efficiency of estuaries (30).

It would be expected that the composition of estuarine waters could also be affected by the flushing velocity, the rate of addition of stream water. Differences in drainage basins produce the major differences in composition between their respective estuaries, by controlling the composition of the incoming water and the rate of discharge (47). Unfortunately, the variation of river-water composition with streamflow and season almost matches the variation between basins. This flow-rate dependence appears to be the key control of changes, with the exceptions of a few trace elements and the major biological elements. For example, a simple mixing formula

$$X_{river} = \frac{X_{sample} \, Cl_{sea} - X_{sea} \, Cl_{sample}}{Cl_{sea} - Cl_{sample}}$$

has good predictive capability for the standard water-quality ions (Na^+, K^+, Ca^{2+}, Mg^{2+}, Cl^-, SO^{2-}_4, HCO^-_3), and provides a comparison between observed concentrations in an estuary in terms of a "river-water concentration" with that in the river-water source (47). Measurement of chlorinity makes it possible to compute the effects of dilution with sea water, and to estimate the effects of other processes. The results are not as clear as they might be because of (a) transformation of the constituent into a form not determined in the analytical procedure, (b) loss to sediments, or (c) variability, with time, of the flow rate of the river. We know that (c) exists, and it may mean that the water in the more seaward part of the estuary has passed the point of contaminant input at a high rate of flow and never had the concentration that was being produced at the source point at the time of the survey (47).

One of the key effects of mixing is the change in pH, where acid river waters having a pH of 6 or less meet the ocean water, which has a pH of 8. The effect is opposite to that for the addition of ground water to stream water, and can lead to strongly increased adsorption.

Ice Transport

The annual load delivered by ice to the oceans has varied with geologic time, as periods of glaciation waxed and waned. The Scandinavian and Canadian shields are areas where such movement has been quantitatively important in recent geologic time. Estimates of the earth's surface covered by ice today range from about 15.1×10^6 km^2 to 16.7×10^6 km^2 (12, 50). Estimates for the total surface covered by the maximum extent of glaciation are about 50×10^6 km^2, or three times larger. Of the area covered today, 85 percent is in the South Polar region (14).

The degree to which ice can erode rock is open to question, but there is no question about the effectiveness of a glacier's ability to move material. The mechanisms of ice movement are complex (see, for example, Ref. 34, p. 149f), and determining the concentration of sediment in ice is even more difficult than that in water. Mixing seems to be minimal, so average values are questionable. Estimates of the amount of sediment transported by ice vary greatly—from $2-5 \times 10^8$ tonnes based on sedimentation rates (51), to 20×10^8 tonnes (11), to $350-500 \times 10^8$ tonnes (50). Major problems in determining this value concern the amount of ice delivered, sediment concentrations, and sediment density. Distribution of the sediments depends on the melting pattern of the ice. For pollution considerations, the Antarctic sediment load must be considered minor—even though the amount of the sediment is a significant proportion of river-delivered sediments (270×10^8 tonnes). The Arctic sediment load is also low, but locally ice-rafted material can contribute more than 30 percent of the $2-3$ mm/10^3 yr of sediment (52).

Transport by the Atmosphere

Land-Atmosphere Interface

Loss of chemicals from surfaces by vaporization, followed by movement into the atmosphere (termed volatilization), is a general function of the vapor pressure of the chemical and its rate of diffusion through the air surrounding the substance. Several transfer agents account for such losses, such as the volatile hydrocarbons that may escape from oil and gas reservoirs and pass into the air through the soil; recent work has focused on this process as a technique for geochemical exploration. Second, the radiogenic gases radon, helium, argon, krypton, and xenon can be produced by the decay of naturally occurring elements; the presence of radon has been used in prospecting for uranium. A third example is the volatilization of rocks and minerals, but the only element that has been extensively studied is mercury (Hg), which can move significant distances. Compounds such as H_2S are thought to be produced by ore-forming processes, and are being studied as tools for geochemical exploration (53, 54, 55).

A fourth naturally occurring process is exudation by plants during the transpiration process. The relative enrichment of elements in soil, plant, and exudates collected over conifer trees suggests the possibility that the elements are carried with the organic constituents, predominantly terpenes (56). Terpenes can be activated photochemically and agglomerated into aerosol particles as large as 2 microns, creating the blue atmospheric haze that is common in forested areas. These metal-bearing aerosols may remain in the atmosphere for several hours before they precipitate. Grasses and small herbaceous plants also release metallic elements to the atmosphere (57). Submicrometer-sized particles are probably released under all conditions, but are at a maximum during times of high transpiration.

Fifth, mechanical disturbance of the plant (e.g., wind shaking) increases the size of the particles released. While larger particles would be expected to settle out quickly, the smaller ones could be transported and dispersed over large areas and thus result in the unexplained presence of metal-rich particles in remote regions. This behavior has also been used as a geochemical exploratory tool.

It is interesting to note that elements can be taken out of the atmosphere and added directly to plants. Analyses of Spanish Moss (58) have shown that 4 of 6 samples containing detectable amounts of Sn were collected within 50 miles of the only tin smelter in the United States — indicating that Spanish Moss might be useful in detecting local atmospheric metal pollution.

Sixth, natural fires (started by lightning) in forest, bush, and grass release large quantities of smoke and trace gases into the atmosphere. Those elements that are present in the plant but not concentrated in the ash could accompany the particles and gases produced. As mentioned in Chapter 6, the transport varies with species and soil. The volume is about 4×10^6 t/day (about 1.5×10^8 t/yr), although human-induced fires are included (59) (Table 3-5). For comparison, industrial particulate pollution in the United States is estimated to be 5–13 times greater, or 20.6×10^6 t/yr (60).

Seventh, wind erosion of soil is generally considered to be less important than water erosion, but it can be locally devastating in dry regions. Wind is estimated to be the dominant cause of erosion in about 3 percent of the total land area of the United States, but it is estimated to be the cause of about 25 percent of the total erosion in areas subject to drought. Vegetation cover is the most important protection against the effect of wind, by reducing wind speed close to the ground and by preventing the direct wind force from reaching erodible particles. Wind erosion occurs not only in the Great Plains but also in the Northwest, Great Lakes, and coastal areas. Those parts of the United States more susceptible to wind erosion than all types of erosion are indicated

TABLE 3-5
Natural Production of Atmospheric Particles (Million Tons/Year)

Source	Quickly Removed (> 5 μm)	Possibly Transported and Dispersed (< 5 μm)	Remarks
Sea-salt aerosols	300-500	300-500	Distribution is assumed equal, uncertainty is 10X.
Wind-blown dust	130-1,000	130-1,000	Uncertainty is 100X.
Forest fires	30-130	5-20	Estimated from industrial smoke, varies greatly with time, uncertainty is 10X.
Volcanoes	150	15-30	Highly variable, 10X uncertainty.
Atmospheric gas aerosols (converted)	100	500	70 percent sulfates, 15 percent hydrocarbon, 15 percent nitrates, uncertainty is 10X.
	710-1,880	950-2,050	

Source: Refs. 50, 61, 62, 63. Estimated direct and converted particle emissions from all countries due to human activities in 1968 were estimated to be 104,000,000 and 25,000,000 t/yr for particles greater than 5 μm. and 30,000,000 and 250,000,000 t/yr for particles less than 5μm. Thus, human activity can contribute 15 to 25 percent of the total yearly amount of dispersed materials (64).

in Figure 3-6. Soil loss equations for wind erosion (61) are available, but, like those for water erosion (42), are useful locally, not regionally. Estimates of U.S. loss of such wind-blown sediment are about 1 billion tons each year; indeed, one estimate shows 800 million t in one year in the western region of the United States alone.

The particulate load of the atmosphere is calculated as 13×10^6 tonnes by assuming a mean dust content of 5×10^6 g/m³, atmospheric thickness of 5000 m, and the Earth's surface area of 510×10^2 m² (62). If the atmospheric sediment is washed out about 40 times per year, the total material added to the atmosphere is about 5×10^8 tonnes (5×10^{14} g). This is about half of the total amount of wind erosion estimated for the United States alone. Another estimate is 1.3×10^8 tonnes, based on calculations from arid zones in the Middle East and extrapolated for the world (63). Despite the definite nature of these numbers, we really do not know the magnitude of the flux, and these values could be low by a factor of 100. Clearly, the bulk of the wind-blown sediment is deposited quickly and locally, but sediment-yield information analogous to that for rivers has not been acquired. As Whelpdale and Munn stated (63, p. 293), "The scientific community has a considerable need for better estimates, particularly of the natural sources. Global models of air pollution are, in fact, often difficult to verify simply because of inaccurate source inventories." This statement has been reinforced by a recent study by the National Academy of Sciences (67).

Recent analysis of atmospheric particulate matter has focused on determination of the amount that is added through pollution. Two major techniques have been used. The first uses a ratio technique for determining abnormal elemental enrichment. For example,

$$\frac{(X/R)_{\text{sample}}}{(X/R)_{\text{crust}}}$$

where X is the element of interest and R is a reference element assumed to be well balanced and not greatly affected by the process. Elements used have included aluminum (68, 69, 70, 71), iron (72, 73), and scandium (74). The assumption is that most atmospheric dust is derived from soil erosion, and that the crustal rock average is a good indication of the composition of material that is removed. Any ratio near unity in Table 3-6 should indicate that crustal material is the probable source, although the variation in soil/crust enrichment ratios would disguise any variation smaller than 100. The problem of using soil source for the particles instead of "average rock" is recognized, and a recent work (75) tried to deal with it. The uncertainties involved in the numbers limits any comparison to only those differing by orders of magnitude.

The second major technique is multivariant factor analysis of the results of chemical measurements. One effort (68) identified the particulate proportion added by natural dust, sea spray, or pollution. Another (76) illustrated that the soil chemistry of Arizona can account for 50 percent of the chemical variation of air particles. One interesting aspect is that nickel and zinc, commonly considered as added through pollution, are related to the soil factor in the Arizona experiment.

The eighth, and last, of the land-air transfer agents to be considered is volcanoes. Volcanic release of both particulate matter (ash) and gases can be considerable (Table 3-6) and, because they can be injected directly into the stratosphere, residence times can be long. About 150×10^6 tonnes (2 percent of the total volcanic release) is added to the stratosphere, with the weight of the particles 10 times the weight of the gases (59, 77). The 1974 Fuego volcano in Guatemala alone contributed at least 4×10^5 tonnes of sulfur and chlorine to

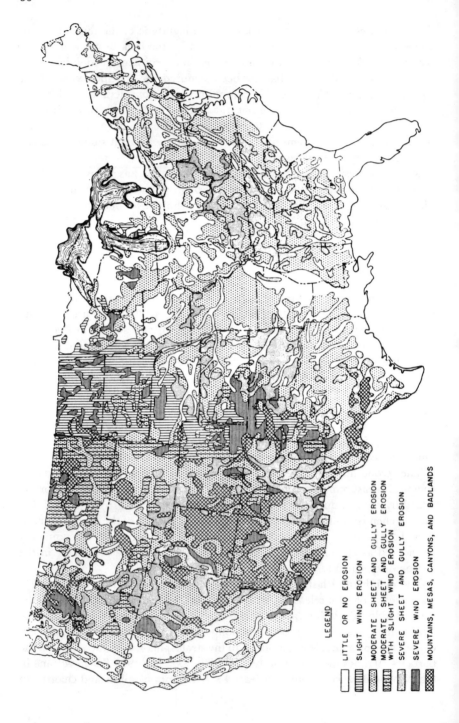

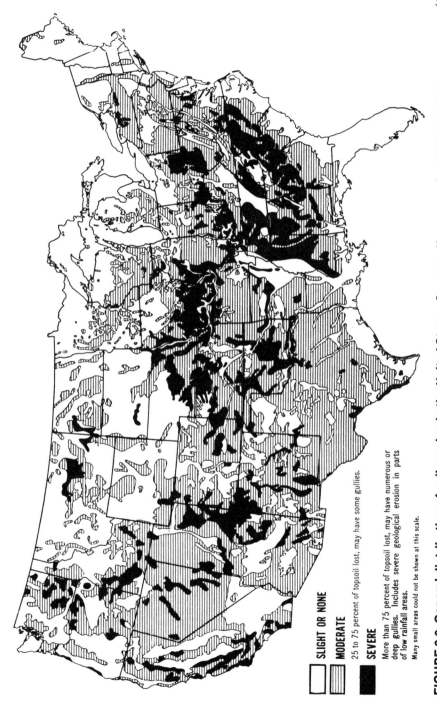

FIGURE 3-6. General distribution of soil erosion in the United States. Sources: Upper map for 1937, Ref. 65; lower map for 1955, Ref. 66.

☐ SLIGHT OR NONE
▥ MODERATE
25 to 75 percent of topsoil lost, may have some gullies.
■ SEVERE
More than 75 percent of topsoil lost, may have numerous or deep gullies. Includes severe geological erosion in parts of low rainfall areas.
Many small areas could not be shown at this scale.

TABLE 3-6
Atmospheric Trace Element Enrichment Coefficient (EF)
$EF_{crust} = (X/R)_{air} / (X/R)_{crust}$

	Oceanic atmosphere				Air near volcanoes			Urban Air
	Ref.68 Bermuda 1973	Ref.69 N.Atlantic Westerlies	Ref.70 S.Pole	Ref.72	Ref.68 Hawaii	Ref.71 Iceland site		Ref.73
Element	R=Al	R=Al	R=Al	R=Fe	R=Al	(A)R=Al	(B)R=Al	R=Fe
Mn	0.73	1.9	1.4	0.9	2.5		3.5	6
Fe	.97	1.1	2.1	2.5	2.9		1.8	1.0
Al	1.0	1.0	1.0	1.0	1.0	1.0	1.0	.5
Sc	1.1	1.1	.8		1.5	.88	.81	1.0
Ni	1.1			.4-3.5	23			12
Cr	1.7		6.9	.5-3.4	27			11
Co	1.8	3.5	4.7	.4-1.1	4.0		3.3	2
Th	2.1		.9		.7	.27	.20	
Eu	2.5		1.9		3.8	2.7	.9	
Ce	3.6		4.4					
Cu	9.6	84	93	1.1-7.6	100		53	83
Zn	26	40	69	3.2-34	140			270
As	22							310
Ag	52				360			
Hg	65				310,000			1,100
Pb	170	2,300	2,500	10-166	100		115	2,300
Sb	180	3,000	1,300		.4			2,800
Cd	570	300	18,000		320			1,900
Se	2,600	16,000			56,000		21,600	2,500

the atmosphere (78). These elements quickly form acids, which can attach themselves to ash particles and leach soluble elements from the silicate glass and minerals therein. While the larger particles are expected to settle quickly, the smaller ones could be transported for long distances, as shown by the brilliant sunsets that follow volcanic eruptions. The particle size decreases to a few tenths of a micrometer during the several weeks necessary for mixing of the particles in the atmosphere.

The overall effect of volcanic addition of aerosols to the atmosphere can be a cooling one for most of the atmosphere and the surface, because sulfuric acid is highly reflective of solar radiation, which is absorbed by the earth-atmosphere system (79). Silicate dust and ammonium-sulfate particles likewise have a cooling effect amounting to tenths of a degree. The temperature in the stratosphere is raised by the aerosols because they absorb thermal radiation from below more effectively than they radiate it into space. Only a small amount of energy is involved, but the effect reaches several degrees of temperature because of the low density of the atmosphere.

The size distribution of the aerosols is critical because, if larger, they could create a greenhouse effect by blocking radiation from the surface and the troposphere. It is believed that this is what has happened on Venus where the surface temperature is several hundred degrees Celsius. Recent measurements indicate that aerosols in the lower troposphere create an increase rather than a decrease in temperature (80, 81). If increases in both carbon dioxide and particulate matter can cause a warming trend, we might be severely underestimating the rapidity with which such a crisis awaits us. Clearly, better understanding of aerosols produced by volcanoes can be critical in predicting climatic variations.

Finally, we must consider elements added to the atmosphere by volcanic processes (Table 3-6). One would expect that, with molten rock as the source, the more volatile elements would escape with the ash and gas, and that any variation in magma type would not alter the general pattern. Thus, the natural processes of a volcano would tend to enrich the atmosphere relative to the crystallizing rock in the volatile elements, just as anthropogenic processes do. Indeed, the patterns and values of enrichment of volcanic aerosols are comparable to urban air compositions as shown in Table 3-6. This presents a problem: if two processes (volcanoes and humans) produce similar results, how can the effects of one (human-caused pollution) be evaluated? One clue might be in the distribution. Most of the land, and therefore most of the human-caused pollution, is in the Northern Hemisphere. On the other hand, similar compositional results have been found in the Southern Hemisphere, where there are a substantial number of active volcanoes but far less human-caused pollution. Therefore both processes are operating.

Ocean-Atmosphere Interface

The largest single source of material added to the atmosphere is sea-spray aerosols (Table 3-5), amounting to about 1×10^9 tonnes/yr, divided equally between sizes greater and less than 5 μm. Enrichment factors for an element relative to sea water, similar to factors discussed previously, are:

$$EF_{sea} = \frac{X/Na_{air}}{X/Na_{sea}}$$

Sodium is almost always used to normalize the analyses because the other obvious choice, chlorine, is highly reactive in sea spray and is quickly converted to gaseous chloride compounds. One study (82) used the chlorine ratio and found that the concentration ratio in the marine atmosphere is proportional to the 0.67-power of the concentration ratio in the ocean. This was attributed to the fact that the surface of the ocean is the physical source, and the surface-to-volume ratio is a 0.67-power function. The ocean appears to be the major source of Na, Mg, Ca, and K in marine aerosols, but can also contribute significantly to the abundance of other elements. Indeed, there is some evidence (75) for significant fluxes from sea surface to atmosphere for Hg, Se, and As.

Sources should not be confused with enrichment; however, Mg, Ca, and K (and Sr) are not enriched in marine aerosols relative to sea water when the atmospheric dust component is less than 5 percent—implying that, in cases where they are enriched, the dust component has caused the enrichment (82, 83) and strong geographic control of Ca, K, and SO_4 indicates continental loading (84).

It appears that Fe, Zn, and Cu are fractionated in sea-salt particles, produced by bubbles bursting at the ocean-air interface. There is very little question that the organic matter on atmospheric particles is highly enriched relative to the sea-water source of such matter, apparently originating from the dissolved or colloidal fraction of the organic matter in sea water (68). The apparent concentration of organic matter in ocean rain is about 8 g/l, considerably greater than that in the ocean (4 mg/l). This creates no problem, however, because surface concentrations of 1 mg/m^2 are common, and a thickness of only 0.1 μm is required to give the higher (8 g/l) ratio (85).

The sea-surface boundary is not sharp, but, as indicated by continuous changes in refractive-index density and dielectric constant from the atmosphere to sea water, possesses properties and composition that justify separate treatment. The transfer from laminar to turbulent flow at about 3 mm ocean depth has been suggested as an appropriate boundary (85).

Continentally derived, wind-carried material appears to account for most of the enrichment in Fe, Al, Mn, Ce, Cr, Ni, Co, Sc, Th, and Eu, and (as mentioned earlier) possibly Mg, Ca, K and Sr—with the sources for Pb, Zn, Cu, Hg, Cd, Se, As, Sb, and Ag still in question (as mentioned in the section on volcanoes). Deposition of this wind-derived sediment is not evenly distributed in the ocean, but is governed according to zonal wind patterns. The geometric mean mineral-aerosol concentrations are generally low over open oceans: 0.36 µg/m^3 in the Pacific between 30°N and 40°S. And they are generally high in trade winds area off continental deserts: 14.2 µg/m^3 in the tropical North Atlantic, 4.3 µg/m^3 in the Mediterranean Sea, and 4.8 µg/m^3 in the Indian Ocean (86).

Atmosphere Removal and Residence Times

The major processes for removal of substances from the troposphere are (a) escape into the stratosphere (gases); (b) wet removal by precipitation (gases and particles); (c) dry removal by sedimentation (particles); (d) adsorption at land and ocean surfaces, including vegetation surfaces (particles and gases); and (e) chemical reaction to produce aerosols and/or absorption on aerosols with subsequent removal (gases). The exchange time of air between the stratosphere and troposphere is nearly one year (87), but it could be only several months (88). The reverse time, from troposphere to stratosphere, should be four years, because there is four times the amount of air in the troposphere as in the stratosphere. Only if other sink mechanisms are very slow (residence times of many years) would the magnitude of the stratosphere sink be expected to be major; the tropopause (at about 10 km) can be considered a major boundary.

Wet removal by precipitation may be incorporated into clouds (rainout or snowout) or below clouds (washout). Residence times of water in the atmosphere are less than 10 days, as discussed previously; 40 rainfalls a year would sweep out all materials below a height of 5 km above the Earth's surface (77). Enough rain analyses are not available and, in any event, differentiation between process (b) and (e) is difficult. Wet removal of gases is considered to be a function of gas solubility. A general term for the removal rate is given by

$$T \approx \frac{8000 \text{ (yr)}}{\alpha}$$

where α is the solubility coefficient, obtained by dividing saturation-vapor density by solubility (87). For most pesticides and toxic chemicals the residence time would be longer than one year if no absorption or aerosols occurred. Where the evaporation rate is greater than that of water, looping will reinforce and prolong the solubility coefficient. Reactions such as oxidation of H_2S and SO_2 followed by aerosol formation are assumed to be much more efficient, and are the principal removal path for sulfur and nitrogen compounds and hydrocarbons. Small aerosols formed initially will coagulate to form larger particles that are readily incorporated into cloud, fog, and rain, and then removed by precipitation (89). Estimates of aerosol-residence times are about 5 days in the lower troposphere and 10–15 days in the upper (87), with an average of about 7 days. (Compare this to the residence time in water of 9.2 days and the assumption of complete washout every 9.1 days.) There is some evidence that there is no vertical difference in the troposphere with the total time less than one week (88). It is important to know the ratio of the amount of gaseous material that becomes attached to aerosols to the amount that remains in the gaseous phase, because of the major difference in residence times between those two processes. Some preliminary attempts to determine this on the basis of absorption theory have been made (87), but information remains scarce. The aerosol concentrations themselves seem to be symmetrically distributed about the Earth (90).

Acid precipitation in eastern North America has been perceived to be a major problem (91). The concentric pattern illustrated in Figure 3-7 originated after the 1920s, with highly acid precipitation (pH less than 4.5) first observed in 1949. The patterns appear to be spreading and the acid content increasing, presumably because of increased solutions of the acids H_2SO_4, HCl, and HNO_3. There is good agreement with forest-covered areas. Apparently less natural dust is produced there than in prairie areas, thus decreasing the possibility of neutralization of the acid by alkaline particles (92a). Since sulfates appear to be spread equally through atmospheric dispersion in industrial areas, the pattern is controlled by dust availability. There is no question the addition of dust, lime, and ash can locally raise pH above 5.6, which is the value of distilled water in equilibrium with atmospheric CO_2. It is questionable, however, as to how widespread the process is, and, indeed, how clearly the data show such trends (92b). Similar increases in acidity appear to occur in Northern Europe; in some small areas the yearly average pH was below 4 (93). In Scandinavia, H_2SO_4 seems to be dominant. In the United States, HNO_3 is a major constituent also. The reasons for the increase in HNO_3 and the role of the other major mineral acid, HCl, are not known (94). Certainly, NH_3 and other bases will neutralize H_2SO_4, reduce the acidity of precipitates, and form particles of $(NH_4)_2SO_4$ and NH_4HSO_4, as previously discussed. The pH will increase but the effect on soils and surface waters will be the same because $(NH_4)_2SO_4$ will have its ammonium exchanged in plants for H.

Changes in surface water show variations that are random, seasonal, yearly, and long-term. Seasonal variations are related to melting snow, with most of the ions entering in the first liquid, so meltwater is more salty and acidic than snow. In the seasonal stage, through underground flow, pure water displaces alkaline ground water, which quickly restores the pH when it

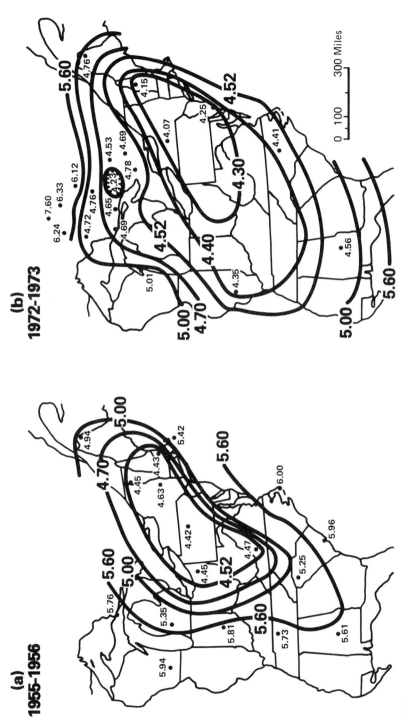

FIGURE 3-7. Distribution of acid rain in northeastern United States in 1955–1956 and 1972–1973. Source: Ref. 91. Published with permission of D. Reidel Publishing Co.

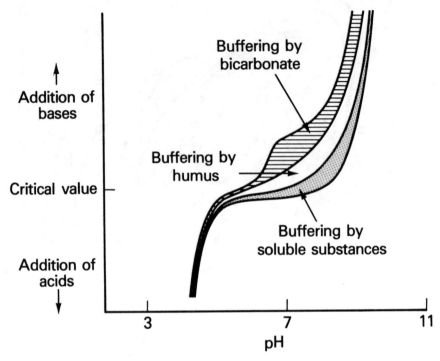

FIGURE 3-8. Buffering patterns of natural waters. Source: Ref. 93.

reaches the surface. Long-term trends can be fitted with regression lines and used to predict the "lifetime of health" of rivers by extending the curves to the time when pH of 5.5—the biologically critical value (93)—is reached. In Sweden, 90 percent of the rivers reach that value in 80 yr.

The buffering capacity of water is a measure of its ability to resist change across a variety of compositions. For example, the pH of water is buffered by soluble materials (phosphates, dissolved organics); colloidal organic material (humus); and bicarbonate. As Figure 3-8 indicates, there is a wide range of compositions where little change in pH occurs, but the shapes of the curves strongly show that very rapid pH changes from 6.5 to below 4.5 can occur with only minor additions of acid if the critical value is exceeded. Odén (93) considered that value to be reached when bicarbonate disappeared from the water. A plot of HCO_3/SO_4 should also show a decrease of the stabilizing effect of bicarbonate content. If the trends continue, HCO_3 disappears, the pH is about 5.5 (the critical value), and the projected time is about the same as for the pH projections (93).

The pH changes in the hydrologic cycle are indicated in Figure 3-9. An increase in soil acidity can lead to an exchange of cations on the soil colloids, a

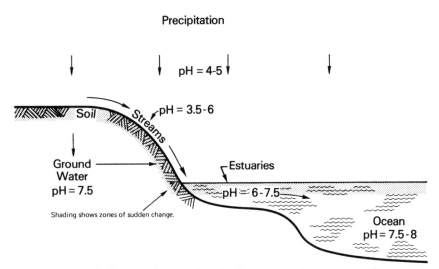

FIGURE 3-9. pH changes in geochemical movement processes. Sources: Ref. 93 and others.

leaching of the ions, and a decrease in the degree of gas saturation; this trend will extend downward in the soil with time. When an accumulating horizon is affected, large amounts of previously held heavy metals may be mobilized, perhaps to be deposited again downstream where pH is increased. It can also affect the biogeochemical cycle by stripping the humus of previously retained elements. There is strong evidence of the addition of sulfur to the soil through acid rain (95), possibly followed by biogenic volatilization and followed again by sulfate deposition. The degree of accumulation is not yet evident.

Dry deposition removes roughly 25 percent of the amount of material carried down by precipitation (96), and is an effective process only for particles larger than 20 mm in diameter. The fall velocity of smaller particles is less than atmospheric vertical movement. Thus, they tend to remain suspended until carried out by precipitation, and their distribution is a function of atmospheric motion.

Four genetic components of this motion can be examined: local, zonal, interzonal, and global. The local one is expected to contain most of the material, analogous to the sediment-delivery rate of the streams. Generally, the mineralogy of the particles varies, as do the soils of their source land areas. This leads to the following zonal relationships: (a) the particular distribution is mainly confined to one wind system (trade, westerlies, etc.); (b) the solids are similar to soil and rock in the adjacent land masses, and probably originate there; and (c) the compositions are indicative of these source areas, which themselves may include an anthropogenic component. There is some transfer between systems because no abrupt transition exists. The interzonal component thus transfers between systems, becomes mixed with zonal particles, and decreases in concentration away from the boundary of the two wind systems. One reason for the lack of widespread mixing of particles in the troposphere is the relatively short residence time, mentioned earlier. There is some evidence from fall-out information that approximately 5 percent of the total soil-sized particles have a residence time of about 100 days, can be transported from hemisphere to hemisphere in that time (especially if introduced into the stratosphere), and have an average composition probably the same as the interzonal values. It is obviously difficult to distinguish among the components of these particles. Some recent work has indicated that soil dust from Asia is a major contributor to marine sedimentation rates in the North Pacific (97). Variation of deposition with storms over the possible source area is needed to show temporal variations.

There is also a background atmospheric particulate matter, composed possibly of $(NH_4)_2SO_4$. At least 50 percent of the particles less than 0.26 μm (and possibly less than 0.55 μm in diameter) consist of a sulfate. This background can also explain the factor-analyses determination of $(NH_4)_2SO_4$, mentioned previously (76). It is appropriate to note that the map of the small-particle concentration (98) correlates high concentrations of background particles in regions where evaporation greatly exceeds precipitation (99) — that is, where washout would be expected to be lowest. There is some evidence that these particles are concentrated in the 0.10-15 μm-size range, with a separate frequency peak around 10 μm for those particles of a transient nature.

The preceding discussion emphasizes the futility of trying to generalize about the composition of rain. It varies drastically from ocean to coast to inland, with the season, and with solid content, among other factors. It is almost impossible to distinguish the different sources of elements present in rain water, yet such must be done to evaluate accurately the different processes. These processes are very rapid — certainly the most rapid of any discussed so far — and are very effective in removing particles and gases. We have very little analytical information on which to base our speculations.

Sorbtion

A recent major review (100) emphasized that to discuss the geochemical movement of elements in rivers it is necessary to know the proportion of movement that takes place in solution and the proportion that takes place associated with the movement of sediments. From a public health perspective, the elements in solution or carried as soluble complexed colloids are readily mobile for incorporation into living material, and therefore comprise an immediate problem. From a practical view, movement of elements associated with the sediments dominates total movement, as indicated in Figure 3-10 (29).

Thus, how elements are carried by sediments and how they might be released into solution together become key problems in evaluating element availability. Because the episodic event appears to transport most of the sediment, study of river-bottom sediments, at least in temperate climates, may give a more accurate picture of chemical transport on the scale of scores of years than does the study of suspended sediments. This expectation has been used for years by exploration geochemists and provided the base for the recent geochemical survey of England and Wales (101).

Even those materials in solution do not necessarily remain there. For example, riverborne dissolved iron is almost completely composed of a mixed oxide–organic colloid that is stabilized by dissolved organic matter (102). When the solution reaches the ocean, flocculation occurs as the sea water neutralizes the negatively charged colloids; the resulting precipitation reduces the input of iron to the ocean to about 10 percent of that in the river, strongly affecting mass-balance and residence-time calculations. Not all elements appear to be rapidly removed from solution upon reaching the estuary. For example, on the basis of geochemical balances, Mn, Ni, Cu, Zn, Cd, and Pb appear to be mobile within the Chesapeake Bay (103). On the other hand, work in the Elbe River (104) shows a decrease in concentration of metals in both suspended sediment and water by the time the ocean is reached. The elements appear to attach onto anything that is present, resulting in total transfer from the stream into the sediments when the estuary is reached.

There is no question that biota quickly pick up elements that are in solution. This is true to such an extent that it has recently been suggested that algae should be grown to act as a control on heavy-metal availability (105).

The chemical and physical nature of solids in the stream controls the availability of the elements. These solids have been divided into a variety of categories: an ion exchange phase, an organic phase, a metal-oxide coating that is sometimes divided into an iron-oxide phase and a manganese-oxide phase (106), and a crystalline phase that is sometimes divided into sulfides, carbonates, and silicates (107, 108, 109). The ion-exchange capacity has commonly been coupled with the presence of clay-sized particles but, as indicated in Figure 3-10, this is a relatively minor transport phase. Indeed, the clay-sized material appears to be important because such particles act as a mechanical substrate for the oxide and organic phases rather than as a major carrier in their own right (107). The organic matter is generally composed of fragmented and indigenous biota, such as algae slimes. Transport is a function of river regime and local vegetation, and is highly variable. As different organisms concentrate different elements in their different parts, the load transported is controlled by the distribution and type of biota (see Chapter 6).

It appears that, with the exception of carbonates and possibly some sulfides, it is not the solubility of the phase itself that controls the availability of

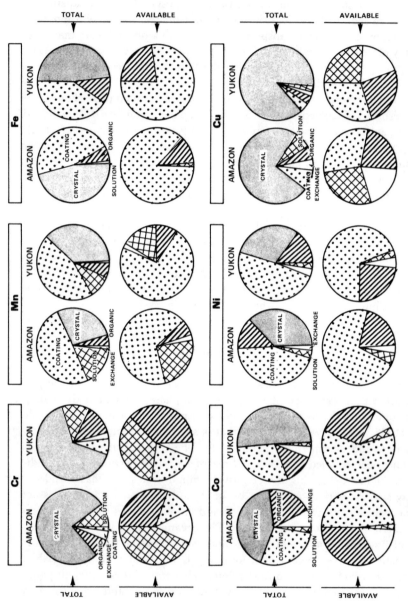

FIGURE 3-10. Transport of some transition elements in the Amazon and Yukon rivers. Available distribution is for other than crystalline material. Source: Modified from Ref. 29. Published with permission of the Geological Society of America and the author.

elements. Instead, surface chemical controls or sorbtion can be used to account for the removal (or addition) of particular elements.

Solids placed in an electrolyte such as water develop charged surfaces, which determine many of their functions, including the extent of adsorption. Through adsorption, the charged suspended solids can modify the composition of water, its pH, and its trace-element load. Where the surface charge is negative the solids are cation exchangers, and where positive they are anion exchangers. Where no charge is present, there is little exchange capacity (110). This surface charge is pH dependent—negative at high pH (basic or alkaline chemical environment) and positive at low pH (acid environment). The value of the pH where the surface charge is neutral is of great interest, because anion- and cation-exchange capacities are equal and minimal, whereas coagulation and sedimentation rates are maximum. Unfortunately, surface charge is not a reliable guide to predicting behavior (111). For example, all solids, regardless of inherited surface charge, become negatively charged in sea water. Thus, the complexing capability of the medium must be considered, and changes can be expected at each reservoir change.

The observed uptake of hydrolyzable metal ions follows the general patterns (112) indicated in Figure 3-11 (113, 114, 115): (a) uptake is strongly dependent on pH and usually occurs over a narrow pH range; (b) the pH value is generally characteristic of the metal ion and its complexes, and is relatively insensitive to the adsorbent (unless the metal ion is a manganese oxide); (c) the pH for a particular metal can be predicted from the hydrolysis curve where the adsorption curve is several pH units lower; (d) H is released as adsorption progresses—that is, the surface of the adsorbent tends to reduce the ion's charge and move to a lower pH; and (e) the amount of adsorption is dependent on the amount of element present in solution, but can usually be expressed as a relatively simple relationship (for example, Ref. 116). This pattern can be used to explain why Zn deficiencies occur at pH of 6 or higher (117)—that is, the Zn is adsorbed, and therefore not available. Other experiments (118) show that Cd follows Zn closely, with increasing adsorption as pH increases in the range of 5 to 7.5. Lead adsorption appears to be 1 pH unit lower than Cd, and Zn and Tl appear to be several units higher. The adsorption of Zn on manganese oxides occurs 2 pH units lower, and that for Pb is 4 units lower, with most of the pH-adsorption occurring before the pH reaches as low as 2. The behavior of Mn appears to be exchangeable between a pH of 4.5 to 6.5, with all of the Mn held at the higher pH level (117).

Clearly, a decrease in pH from 6.5 to 4.5 can suddenly release into solution elements that had been held tightly adsorbed to a solid phase at the higher pH. Certainly Tl, Cd, Zn, Mn, and Pb follow this pattern. As pointed out in Figure 3-8, a critical value of buffering capacity appears to exist such that, if exceeded, the system is overloaded and the pH can quickly change by several

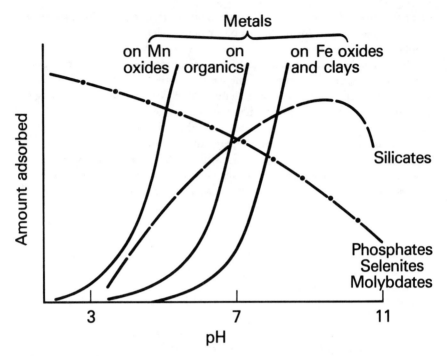

FIGURE 3-11. Generalized model for cation and anion adsorption. Sources: Refs. 113, 114, and 115.

orders of magnitude. Thus, an unanticipated impact of increased runoff acidity because of acid rain may be a sudden surge of heavy metals moving from the solid phase into solution, where they are easily incorporated into biota.

Examination of Figures 3-9 and 3-11 also provides a possible explanation for two other phenomena. First, bottom sediments in streams, especially toward the banks, are almost always covered with dark oxide coatings. This indicates the strong lateral addition of high-pH ground water to streams, with quick precipitation of the oxides carried by the more acidic runoff water. Second, as the river moves into the brackish water of the estuary, the pH rises sharply. Metals are removed from solution and adsorbed on whatever solid phase happens to be available. The estuary is not only a physical barrier to transport from river to ocean, but a chemical one as well. This is shown by the work on the Elbe River (104) and indicated by ongoing work on the Hudson River (119).

The Fe-Mn oxide coatings are not the same. Iron (+2) is generally oxidized and largely precipitated from oxy-generated water at pH greater than 7. The iron that is apparently in solution in river water is actually present as colloidal Fe, or is associated with organic matter (120). Iron present with Mn

TABLE 3-7
Elements Adsorbed on Iron and Manganese Oxides on Soil Particles and Stream Sediments

Strongly Adsorbed		Weakly Adsorbed	Probably Not Adsorbed	Not Adsorbed
By Fe Oxides	By Mn Oxides			
In(+)	Ba	Ag	Be	B
As	Cd	Cu*	Ca	Cr
	C_o	Pb(+)	Ga	K
	Ni*		La	Mg
	Te		Mo	Sc
	Zn		Rb	Ti
			Sb	U
			Sr	Zn
			Y	

Source: Refs. 29, 102, 107, 111, 114, 115, 116, 117, 118, 120, 121, 122, 123, 124.

* not adsorbed by Fe-oxide
(+) not adsorbed by Mn-oxide

oxides at high pH is due to adsorption of the iron hydroxide on the Mn oxide surfaces. Manganese oxides are more important than Fe oxides as a cation-sorbant base — not only because of the much lower pH zero point-charge, but also because Mn can occur in +2, +3, and +4 oxidation states (all of different size) and the cation-sorbant base can accommodate many different metal ions as Mn substitutes. Some elemental behavior is indicated in Table 3-7. Unfortunately, too many studies have considered the Fe and Mn oxides together rather than distinguishing between an iron-oxide phase and a manganese-oxide phase.

There is indication that the physical and chemical characteristics of stream sediments are the same as those of the soil particles of the river basin (49, 125). Certainly the movement of water and that of sediments would support this idea. It seems clear that, to understand fully what is happening in a river, we must understand what is happening in the soil reservoir.

References Cited

1. Benecke, P., 1976, Soil Water Relations and Water Exchange of Forest Ecosystem; p. 101-131 *in* Water and Plant Life, O. L. Lange, L. Kappen, and E. D. Schulze, eds., Springer-Verlag, Berlin, 536 p.

2. Branson, F., 1975, Natural and Modified Plant Communities as Related to Runoff and Sediment Yields; p. 157-172 *in* Coupling of Land and Water Systems, A. D. Hasler, ed., Springer-Verlag, Berlin, 307 p.

3. Bricker, O., 1968, Cations and Silica in Natural Waters: Control by Silicate Minerals; p. 110-119 *in* Geochemistry, Precipitation, Vaporation, Soil-Moisture, Hydrometry. Publication of the International Association of Scientific Hydrology, no. 78, Gentbrugge, Belgium, 431 p.

4. Meybeck, M., 1976, Total Mineral Dissolved Transport by World Major Rivers; Hydrological Science Bulletin, v. 21, no. 6, p. 265-284.

5. Meybeck, M., 1977, Dissolved and Suspended Matter Carried by Rivers: Composition, Time and Space Variation, and World Balance; p. 25-32 *in* Interactions Between Sediments and Freshwater, H. L. Golterman, ed., Dr. W. Junk, B.V. Publ., Hague.

6. Gibbs, R. J., 1967, The Geochemistry of the Amazon River System: Part I. The Factors that Control the Salinity and the Composition and Concentration of the Suspended Solids; Geological Society of America Bulletin, v. 78, p. 1203-1232.

7. Holland, H. D., 1978, The Chemistry of the Atmosphere and Oceans; Wiley, New York, 351 p.

8. Alekin, O. A., and L. V. Brazhnikova, 1968, Dissolved Matter Discharge and Mechanical and Chemical Erosion; p. 35-41 *in* Geochemistry, Precipitation, Vaporation, Soil-Moisture, Hydrometry. Publication of the International Association of Scientific Hydrology, no. 78, Gentbrugge, Belgium, 431 p.

9. Livingstone, D. A., 1963, Chemical Composition of Rivers and Lakes; U.S. Geological Survey Professional Paper 440-G, Washington, D.C., 61 p.

10. Clarke, F. W., 1924, Data of Geochemistry; 5th ed., U.S. Geological Survey Bulletin 770, 841 p.

11. Garrels, R. M., and F. T. Mackenzie, 1971, Evolution of Sedimentary Rocks; W. W. Norton, New York, 397 p.

12. Lvovitch, M. I., 1979, World Water Resources and Their Future; Eng. trans., R. L. Nace, ed., American Geophysical Union, Washington, D.C., 415 p.

13. Martin, J. M., and M. Meybeck, 1978, The Content of Major Elements in the Dissolved and Particulate Load of Rivers; p. 95-110 *in* Biogeochemistry of Estuarine Sediments, UNESCO, Paris.

14. USSR Committee for the International Hydrologic Decade, V. I. Korzoun, ed., 1978, World Water Balance and Water Resources of the Earth, UNESCO, Paris, 663 p.

15. Gibbs, R. J., 1970, Mechanisms Controlling World Water Chemistry; Science, v. 170, p. 1088-1090.

16. Miller, W. R., and J. I. Drever, 1977, Chemical Weathering and Related Controls on Surface Water Chemistry in the Absaroka Mountains, Wyoming; Geochimica et Cosmochimica Acta, v. 41, p. 1693-1702.

17. Uncharted Saltwater Rivers in Peru, EOS, February 27, 1979, p. 143.

18. Hem, J. D., 1972, Water—Nonmarine; p. 1243-1244 *in* Encyclopedia of Geochemistry and Environmental Sciences, R. W. Fairbridge, ed., Van Nostrand Reinhold, New York.

19. White, D. E., J. D. Hem, and G. A. Waring, 1963, Chemical Composition of

Subsurface Water; U.S. Geological Survey Professional Paper 440-F, Washington, D.C., 65 p.

20. Skongstad, M. W., 1971, Minor Elements in Water; p. 43-55 in Environmental Geochemistry in Health & Disease, H. L. Cannon and H. C. Hopps, eds., Geological Society of America Memoir 123, p. 32.

21. Likens, G. E., F. H. Bormann, R. S. Pierce, J. S. Eaton, N. M. Johnson, 1977, Biogeochemistry of a Forested Ecosystem, Springer-Verlag, New York, 146 p.

22. Fournier, F., 1960, Debit Solide des Cours D'Eau. Essai D'estunation de la Perte Enterre Subie Par L'Ensemble du Globe Terrestie; p. 19-25 in Land Erosion, Precipitation, Evaporation, Publication of the International Association of Scientific Hydrology, no. 53, Gentbrugge, Belgium, 527 p.

23. Judson, S., 1968, Erosion of the Land, or What's Happening to Our Continents? American Scientist, v. 56, no. 4, p. 356-374.

24. Holeman, J. N., 1968, The Sediment Yield of Major Rivers of the World; Water Resources Research, v. 4, no. 4, p. 737-747.

25. Jansen, J.M.L., and R. B. Painter, 1974, Predicting Sediment Yield from Climate and Topography; Journal of Hydrology, v. 21, p. 371-380.

26. Langbein, W. B., and S. A. Schumm, 1958, Yield of Sediment in Relation to Mean Annual Precipitation, EOS, v. 39, p. 1076-1084.

27. Wilson, L., 1972, Seasonal Sediment Yield Patterns of U.S. Rivers; Water Resources Research, v. 8, no. 6, p. 1470-1479.

28. Beckinsale, R. P., 1969, River Regimes; p. 455-471 in Water, Earth, and Man, R. J. Chorley, ed., Methuen, London, 588 p.

29. Gibbs, R. J., 1977, Transport Phases of Transition Metals in the Amazon and Yukon Rivers; Geological Society of America Bulletin, v. 88, p. 829-843.

30. Schubel, J. R., 1977, Sediment and the Quality of the Estuarine Environment: Some Observations; p. 399-423 in Fate of Pollutants in the Air and Water Environments. Advances in Environmental Science and Technology, v. 8, part I. I. H. Suffet, ed., Wiley, New York, 484 p.

31. Chen, C. N., 1974, Evaluation and Control of Soil Erosion in Urbanizing Watersheds; p. 161-172 in Proceedings, National Symposium on Urban Rainfall and Runoff and Sediment Control, D.T.Y. Kao, ed., UKY BU 106, University of Kentucky, 246 p.

32. Schumm, S. A., 1977, The Fluvial System; Wiley, New York, 338 p.

33. Robinson, A. R., 1977, Relationship Between Soil Erosion and Sediment Delivery; p. 159-167 in Erosion and Solid Matter Transport in Inland Waters, Publication of the International Association of Scientific Hydrology, no. 122, Gentbrugge, Belgium, 352 p.

34. Statham, I., 1977, Earth Surface Sediment Transport; Clarendon Press, Oxford, 184 p.

35. Karr, J. R., and I. J. Schlosser, 1978, Water Resources and the Land Water Interface; Science, v. 201, p. 229-234.

36. Douglas, I., 1967, Man, Vegetation and the Sediment Yield of Rivers; Nature, v. 215, p. 925-928.

37. Guy, H. P., 1974, An Overview of Urban Sedimentation; p. 149-159 in Proceedings, National Symposium on Urban Rainfall and Runoff and Sediment Control,

D.T.Y. Kao, ed., UKY BU 106, University of Kentucky, 246 p.

38. Meade, R. H., and S. W. Tremble, 1974, Changes in Sediment Loads in Rivers of the Atlantic Drainage of the United States Since 1900; p. 99-104 in Effects of Man on the Interface of the Hydrologic Cycle on the Physical Environment, Publication of the International Association of Scientific Hydrology, no. 113, Gentbrugge, Belgium, 157 p.

39. Schubel, J. R., 1974, Effects of Storm Agnes on Chesapeake Bay; p. 121-132 in Suspended Solids in Water, R. J. Gibbs, ed., Plenum Press, New York, 320 p.

40. Wilson, L., 1973, Variations in Mean Annual Sediment Yield as a Function of Mean Annual Precipitation; American Journal of Science, v. 273, p. 335-349.

41. Wolman, M. G., and J. P. Miller, 1960, Magnitude and Frequence of Forces in Geomorphic Processes; Journal of Geology, v. 68, p. 54-74.

42. Wischmeier, W. H., and D. D. Smith, 1965, Predicting Rainfall-Erosion Losses from Cropland East of the Rocky Mountains; U.S. Department of Agriculture, Agriculture Handbook no. 282, 42 p.

43. De Ploey, J., and J. Savat, 1976, The Differential Impact of Some Soil Loss Factors on Flow, Runoff Creep and Rainwash; Earth Surface Processes, v. 1, p. 151-161.

44. Schubel, J. R., and H. H. Carter, 1976, Suspended Sediment Budget for Chesapeake Bay; p. 48-62 in Estuarine Processes, v. 2, M. Wiley, ed., Academic Press, New York.

45. Meade, R. H., P. L. Sachs, F. T. Manheim, J. C. Hathaway, and D. W. Spencer, 1975, Sources of Suspended Matter in Waters of The Middle Atlantic Bight; Journal of Sedimentary Petrology, v. 45, no. 1, p. 171-188.

46. Drake, D. E., 1976, Suspended Sediment Transport and Mud Deposition on Continental Shelves; p. 127-158 in Marine Sediment Transport and Environmental Management, D. J. Stanley and D.J.P. Swift, eds., Wiley, New York, 602 p.

47. Carpenter, J. H., W. L. Bradford, and V. Grant, 1975, Processes Affecting the Composition of Estuarine Waters; p. 188-214 in Estuarine Research, v. 1, L. E. Cronin, ed., 2nd International Estuarine Resource Conference, Myrtle Beach, S.C., Academic Press, New York, 738 p.

48. Duane, D. B., 1976, Sedimentation and Coastal Engineering: Beaches and Harbors; p. 493-557 in Marine Sediment Transport and Environmental Management, D. J. Stanley and D.J.P. Swift, eds., Wiley, New York, 602 p.

49. Gibbs, R. J., 1977, Suspended Sediment Transport and the Turbidity Maximum; p. 104-109 in Estuaries, Geophysics and the Environment, Geophysics Research Board, NRC-NAS, Washington, D.C., 127 p.

50. Lisitzin, A. P., 1972, Sedimentation in the World Ocean; Society of Economic Paleontologists and Mineralogists Special Publication, no. 17, Tulsa, Okla., 218 p.

51. Hurd, D. C., 1977, The Effect of Glacial Weathering on the Silica Budget of Antarctic Waters; Geochimica et Cosmochimica Acta, v. 41, p. 1213-1222.

52. Mullen, R. E., D. A. Darby, and D. L. Clark, 1972, Significance of Atmospheric Dust and Ice Rafting for Arctic Ocean Sediment; Geological Society of America Bulletin, v. 83, p. 205-212.

53. Hawkes, H. E., and J. S. Webb, 1962, Geochemistry in Mineral Exploration; Harper and Row, New York, 415 p.

54. Siegel, F. R., 1974, Applied Geochemistry; Wiley, New York, 353 p.

54. Siegel, F. R., 1974, Applied Geochemistry; Wiley, New York, 353 p.

55. Rose, A. W., H. E. Hawkes, and J. S. Webb, 1979, Geochemistry in Mineral Exploration, 2nd ed., Academic Press, New York, 657 p.

56. Curtin, G. C., H. D. King, and E. L. Mosier, 1974, Movement of Elements into the Atmosphere from Coniferous Trees in Subalpine Forests of Colorado and Idaho; Journal of Geochemical Exploration, v. 3, p. 245-263.

57. Beauford, W., J. Barber, and A. R. Barringer, 1977, Release of Particles Containing Metals from Vegetation into the Atmosphere; Science, v. 195, p. 571-573.

58. Shacklette, H. T., and J. J. Connor, 1973, Airborne Chemical Elements in Spanish Moss; U.S. Geological Survey Professional Paper 574-E, Washington, D.C., 46 p.

59. Hidy, G. M., and J. R. Brock, 1971, An Assessment of the Global Sources of Tropospheric Aerosols; p. 1088-1097 in Proceedings of the International Clean Air Congress, 2nd, 1970, Academic Press, New York.

60. Waldbott, G. L., 1973, Health Effects of Environmental Pollutants, C. V. Mosby, St. Louis, 316 p.

61. Skidmore, E. L., and N. P. Woodruff, 1968, Wind Erosion Forces in the United States and Their Use in Predicting Soil Loss; Agriculture Handbook No. 346, U.S.D.A., Washington, D.C., 42 p.

62. Goldberg, E. D., 1971, Atmospheric Dust, The Sedimentary Cycle and Man: Comments on Earth Sciences; Geophysics, v. 1, no. 5, p. 117-132.

63. Whelpdale, D. M., and R. E. Munn, 1976, Global Sources, Sinks and Transport of Air Pollution; p. 289-324 in Air Pollution, 3rd ed., A. C. Stern, ed., Academic Press, New York, 715 p.

64. National Academy of Sciences, 1975, Principles for Evaluating Chemicals in the Environment; p. 412, in A Report of the Committee for the Working Conference on Principles of Protocol for Evaluating Chemicals in the Environment, N.A.S., Washington, D.C., 454 p.

65. United States Department of Agriculture, 1938, Yearbook of Agriculture 1938, "Soils and Men," U.S.D.A., Washington, D.C., 1232 p.

66. United States Department of Agriculture, 1957, Yearbook of Agriculture 1955, "Soil," U.S.D.A., Washington, D.C., 784 p.

67. National Academy of Sciences, 1978, The Tropospheric Transport of Pollutants and Other Substances to the Oceans; NRC Workshop on Tropospheric Transport of Pollutants to the Ocean Steering Committee, 1975, Washington, D.C., 243 p.

68. Duce, R. A., G. L. Hoffman, B. J. Ray, I. S. Fletcher, G. T. Wallace, J. L. Fasching, S. R. Piotrowicz, P. R. Walsh, E. J. Hoffman, J. M. Miller, and J. L. Heffter, 1976, Trace Metals in the Marine Atmosphere: Sources and Fluxes; p. 77-119 in Marine Pollution Transfer, H. L. Windom and R. A. Duce, eds., Lexington Books, Lexington, Mass., 391 p.

69. Duce, R. A., G. L. Hoffman, J. L. Fasching, and J. L. Moyers, 1974, The Collection and Analysis of Trace Elements in Atmospheric Particulate Matter over the North Atlantic Ocean; World Meteorological Organization Special Environmental Report no. 3, WMO—No. 368, Geneva, p. 370-375.

70. Zoller, W. H., E. S. Gladney, and R. A. Duce, 1974, Atmospheric Concentrations and Sources of Trace Metals at the South Pole; Science, v. 183, p. 198-200.

71. Mroz, E. J., and W. H. Zoller, 1975, Composition of Atmospheric Particulate Matter from the Eruption of Hermacy, Iceland; Science, v. 190, p. 461-464.
72. Chester, R., and J. H. Stoner, 1974, The Distribution of Mn, Fe, Cu, Ni, Co, Ga, Cr, V, Ba, Sr, Sn, Zn, and Pb in some Soil-sized Particulates from the Lower Troposphere Over the World Ocean; Marine Chemistry, v. 2, p. 157-188.
73. IDOE, 1972, Baseline Studies of Pollutants in the Marine Environment and Research Recommendations; E. D. Goldberg, ed., IDOE Baseline Conference, May 24-26, 1972, National Science Foundation, Washington, D.C., 54 p.
74. Rugaini, R. C., H. R. Ralston, and N. Roberts, 1977, Environmental Trace Metal Contamination in Kellogg, Idaho, Near a Lead Smelting Complex; Environmental Science and Technology, v. 11, p. 773-781.
75. Lantzy, R. J., and F. T. Mackenzie, 1979, Atmospheric Trace Metals: Global Cycles and Assessment of Man's Impact; Geochimica et Cosmochimica Acta, v. 43, p. 511-525.
76. Gaarenstroom, P. D., S. P. Perone, and J. L. Moyers, 1977, Application of Pattern Recognition and Factor Analysis for Characterization of Atmospheric Particulate Composition in Southwest Desert Atmosphere; Environmental Science and Technology, v. 11, no. 8, p. 795-800.
77. Goldberg, E. D., 1976, The Health of the Oceans; UNESCO Press, Paris, 172 p.
78. Rose, W. I., Jr., 1977, Scavenging of Volcanic Aerosol by Ash: Atmospheric and Volcanologic Implications; Geology, v. 5, p. 621-624.
79. Hansen, J. E., W.-C. Wang, and A. A. Lacis, 1978, Mount Agung Eruption Provides Test of a Global Climatic Perturbation; Science, v. 199, p. 1065-1068.
80. Idso, S. B., and A. J. Brazel, 1977, Planetary Radiation Balance as a Function of Atmospheric Dust: Climatological Consequences; Science, v. 198, p. 731-733.
81. Idso, S. B., and A. J. Brazel, 1978 (Reply to) Herman, B. M., S. A. Twomey, and D. O. Staley, Atmospheric Dust Climatological Consequences; Science, v. 201, p. 378-379.
82. Korzh, V. D., 1974, Some General Laws Governing the Turnover of Substance Within the Ocean-Atmosphere-Continent-Ocean Cycle; Journal de Recherches Atmospheriques, v. 8, no. 3-4, p. 653-660.
83. Hoffman, E. J., G. L. Hoffman, and R. A. Duce, 1974, Chemical Fractionation of Alkali and Alkaline Earth Metals in Atmospheric Particulate Matter over the North Atlantic; Journal des Recherches Atmospheriques, v. 8, no. 3-4, p. 675-688.
84. Buat-Menard, P., J. Morelli, and R. Chesselet, 1974, Water-Soluble Elements in Atmospheric Particulate Matter over Tropical and Equatorial Atlantic; Journal de Recherches Atmospheriques, v. 8, no. 3-4, p. 661-673.
85. MacIntyre, F., 1974, Chemical Fractionation and Sea-Surface Microlayer Processes; p. 245-295 *in* The Sea, v. 5, Marine Chemistry, E. Goldberg, ed., Wiley, New York, 895 p.
86. Prospero, J. M., 1978, Mineral and Sea Salt Aerosol Concentrations in Various Ocean Regions; EOS, v. 59, no. 4, p. 306.
87. Junge, C. E., 1977, Basic Considerations About Trace Constituents in the Atmosphere as Related to the Fate of Global Pollutants; p. 7-25 *in* Fate of Pollutants in the Air and Water Environments, Advances in Environmental Science and Technology, v. 8, part I, I. H. Suffet, ed., Wiley, New York, 484 p.

88. Martell, E. A., and H. E. Moore, 1974, Tropospheric Aerosol Residence Times: A Critical Review; Journal de Recherches Atmospheriques, v. 8, no. 3-4, p. 903-910.
89. Robinson, E., and E. C. Moser, 1971, Global Gaseous Pollutant Emissions and Removal Mechanisms; Proceedings, International Clean Air Congress, 2nd, Academic Press, New York, p. 1097-1101.
90. Hogan, A. W., and V. A. Mohner, 1979, On the Global Distribution of Aerosols; Science, v. 205, p. 1373-1375.
91. Cogbill, C. V., 1976, The History and Character of Acid Precipitation in Eastern North America; Water, Air, and Soil Pollution, v. 6, p. 407-413.
92a. Winkler, E. M., 1976, Natural Dust and Acid Rain; Water, Air and Soil Pollution, v. 6, p. 295-302.
92b. American Petroleum Institute and Everett and Associates, 1981, A Detailed Analysis of the Scientific Evidence Concerning Acidic Precipitation, Acidic Deposition Appendix, American Petroleum Institute, Washington, D.C. See also M. L. Miller and A. G. Everett, 1981, History and Trends of Atmospheric Nitrate Deposition in the Eastern U.S.A., p. 162-178 *in* Formation and Fate of Atmospheric Nitrates, H. M. Barnes, ed., Environmental Sciences Research Laboratory, U.S. Environmental Protection Agency, Research Triangle Park, N.C.
93. Odén, S., 1976, The Acidity Problem—An Outline of Concepts; Water, Air, and Soil Pollution, v. 6, p. 137-166.
94. Cronan, C. S., W. A. Reiners, R. C. Reynolds, Jr., and G. E. Lang, 1978, Forest Floor Leaching: Contributions from Mineral, Organic, and Carbonic Acids in New Hampshire Subalpine Forests; Science, v. 200, p. 309-311.
95. Shinn, J. H., and S. Lynn, 1979, Do Man-made Sources Affect the Sulfur Cycle of Northeastern States? Environmental Science and Technology, v. 13, p. 1062-1067.
96. Junge, C. E., 1963, Air Chemistry and Radioactivity; Academic Press, New York, 382 p.
97. Duce, R. A., C. K. Unni, B. J. Ray, J. M. Prospero, and J. T. Merrill, 1980, Long-Range Atmospheric Transport of Soil Dust from Asia to Tropical North Pacific: Temporal Variability; Science, v. 209, p. 1522-1524.
98. Junge, C. E., 1972, Our Knowledge of the Physics-Chemistry of Aerosols in the Undisturbed Marine Environment; Journal of Geophysical Research, v. 77, no. 27, p. 5183-5200.
99. Baumgartner, A., and E. Reichel, 1975, The World Water Balance, Mean Annual Global, Continental and Maritime Precipitation, Evaporation and Runoff; trans. Richard Lee, Elsevier, New York, 179 p.
100. Förstner, U., and G.T.W. Wittmann, 1979, Metal Pollution in the Aquatic Environment; Springer-Verlag, Berlin, 486 p.
101. Imperial College of Science and Technology Applied Geochemistry Research Group, 1978, The Wolfson Geochemical Atlas of England and Wales; Oxford University Press, 69 p.
102. Boyle, E. A., J. M. Edmond, and E. R. Sholkovitz, 1977, The Mechanism of Iron Removal in Estuaries; Geochimica et Cosmochimica Acta, v. 41, p. 1313-1324.
103. Holz, G. R., 1976, Trace Element Inventory for the Northern Chesapeake Bay

with Emphasis on the Influence of Man; Geochimica et Cosmochimica Acta, v. 40, p. 573-580.

104. Muller, G., and U. Förstner, 1975, Heavy Metals in Sediments of the Rhine and Elbe Estuaries: Mobilization or Mixing Effect; Environmental Geology, v. 1, p. 33-39.

105. Jennett, J. C., J. M. Hassett, and J. E. Smith, 1980, The Use of Algae to Control Heavy Metals in the Environment; Minerals and the Environment, v. 2, p. 26-31.

106. Suarez, D. L., and D. Langmuir, 1976, Heavy Metal Relationships in a Pennsylvania Soil; Geochimica et Cosmochimica Acta, v. 40, p. 589-598.

107. Jenne, E. A., 1977, Trace Element Sorption by Sediments and Soils—Sites and Processes; p. 425-553 in Molybdenum in the Environment, v. 2, Geochemistry, Cycling, and Industrial Uses of Molybdenum, W. A. Chappell and K. K. Petersen, eds., Marcel Dekker, Inc., New York, 812 p.

108. Förstner, Ulrich, 1977, Metal Concentrations in Freshwater Sediments—Natural Background and Cultural Effects; p. 94-103 in Interactions Between Sediments and Freshwater, H. L. Golterman, ed., Dr. W. Junk, B. V. Publ., Hague.

109. Förstner, U., G. Müller, and P. Stoffers, 1978, Heavy Metal Contamination in Estuarine and Coastal Sediments: Sources, Chemical Association and Diagenetic Effects; p. 49-69 in Biogeochemistry of Estuarine Sediments, UNESCO, Paris, 293 p.

110. Parks, G. A., 1967, Aqueous Surface Chemistry of Oxides and Complex Oxide Minerals; p. 121-160 in Equilibrium Concepts in Natural Water Systems, Advances in Chemistry no. 67, American Chemical Society, Washington, D.C., 344 p.

111. Parks, G. A., 1975, Adsorption in the Marine Environment; p. 241-308 in Chemical Oceanography, v. 1, 2nd ed., J. P. Riley and G. Skorrow, eds., Academic Press, London, 606 p.

112. James, R. O., and M. G. MacNaughton, 1977, The Adsorption of Aqueous Heavy Metals on Inorganic Materials; Geochimica et Cosmochimica Acta, v. 41, p. 1549-1555.

113. Bowden, J. W., M.D.A. Bolland, A. M. Posner, and J. P. Quirk, 1973, Generalized Model for Anion and Cation Adsorption at Oxide Surfaces; Nature, v. 245, p. 81-83.

114. Quirk, J. P., and A. M. Posner, 1975, Trace Element Adsorption by Soil Minerals; p. 95-107 in Trace Elements in Soil-Plant-Animal Systems: D.J.D. Nicholas and A. R. Egan, eds., Academic Press, New York, 417 p.

115. McLaren, R. G., and D.V.G. Crawford, 1973, Studies on Soil Copper, II. The Specific Adsorption of Copper by Soils; Journal of Soil Science, v. 24, no. 4, p. 443-452.

116. Browman, M. G., and G. Chesters, 1977, The Solid-Water Interface: Transfer of Organic Pollutants Across the Solid-Water Interface; p. 49-105 in Fate of Pollutants in the Air and Water Environments, Part 1, Mechanism of Interaction Between Environments and Mathematical Modeling and the Physical Fate of Pollutants, I. H. Suffet, ed., Wiley, New York, 484 p.

117. Jenne, E. A., 1968, Controls on Mr, Fe, Co, Ni, Cu, and Zr Concentrations in Soils and Water: The Significant Role of Hydrous Mu and Fe Oxides; p. 337-387 in Trace Inorganics in Water, R. A. Baker, ed., Advances in Chemistry, no. 73, American Chemical Society, Washington, D.C., 396 p.

118. Rao Gadde, R., and H. A. Laitinin, 1974, Studies of Heavy Metal Adsorption by Hydrous Iron and Manganese Oxides; Analytical Chemistry, v. 46, no. 13, p. 2022-2026.

119. Speidel, D. H., D. Thurber, D. Locke, and N. Coch, Geochemistry of the Sediments of the Hudson River and Estuary; research in progress.

120. Hem, J. D., 1978, Redox Processes at Surfaces of Manganese Oxide and Their Effects on Aqueous Metal Ions; Chemical Geology, v. 21, p. 199-218.

121. Carpenter, R. H., T. A. Pope, and R. L. Smith, 1975, Fe-Mn Oxide Coatings in Stream Sediment Geochemical Surveys; Journal of Geochemical Exploration, v. 4, p. 349-363.

122. Nowlan, G. A., 1976, Concretionary Manganese-Iron Oxides in Streams and Their Usefulness as a Sample Medium for Geochemical Prospecting; Journal of Geochemical Exploration, v. 6, p. 193-210.

123. Whitney, P. R., 1975, Relationship of Manganese-Iron Oxides and Associated Heavy Metals to Grain Size in Stream Sediments; Journal of Geochemical Exploration, v. 4, p. 251-263.

124. Florence, T. M., Trace Metal Species in Fresh Waters; Water Research, v. 11, p. 681-687.

125. Hassett, J., 1980, personal communication (Department of Agronomy, University of Illinois, Urbana).

4
Soil Reservoir

The dust we tread upon was once alive.
— Byron

Introduction

The soil reservoir is a dynamic one, formed by the interaction of air, water, biota, and rock (see Figure 4-1), and constantly changing in volume as material is added, lost, and recycled in response to environmental conditions. It is the medium that provides, with air and water, the basis for human life through the support of food plants, the release of photosynthetic oxygen, and the cleansing of downward-percolating water. It provides the means for reentry of decayed organic material into the geochemical cycle, a sometime sink for pesticides and heavy elements, and the path by which rock enters the weathering cycle and thus renews the soil reservoir. In general, this renewal or replacement time is significantly longer than one human generation. Soil should therefore be considered a nonrenewable resource in the short term. The problem of soil erosion (the flux of soil from the soil reservoir by means of water or air) has been mentioned as a critical problem deserving the immediate attention of Congress (1). The U.S. Department of Agriculture has concentrated enormous time, effort, and funds since the mid-1930s in attempts to define the problem and derive solutions, but with mixed success.

What constitutes the soil reservoir in the biogeochemical cycle? The total amount of land above sea level is 36.6 billion acres. Of this, only about 11 billion acres is directly usable by man (the so-called *ecumene*) with the remainder too cold, too dry, or too mountainous, and thus ordinarily not developing productive soil (see below). The amount of land that is arable, i.e., in which the soil will support row-crop agriculture, is smaller still—about 6 billion acres. The decrease in arable land through erosion and urbanization by an increasing population (flat, temperate climates are a nice place to live) may cause a problem in food production that technology might not be able to counter. Some farms in the United States lose about 12 tonnes of soil per acre

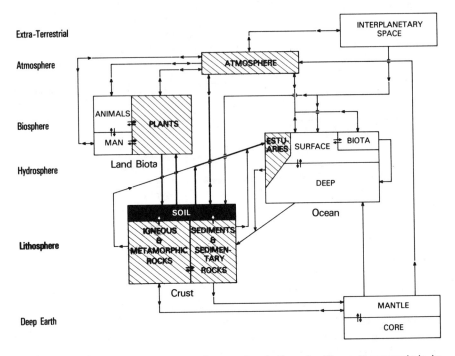

FIGURE 4-1. The soil reservoir in the geochemical cycle. The soil reservoir is in black. Shaded reservoirs indicate those with major interactions with the soil reservoir. Important fluxes are indicated by heavy lines.

per year. The total weight of soil in a plowed section one mile on a side is about 1000 tonnes (or less if the soil has a high organic content), and the time of soil formation is generally hundreds of years. The rate of soil exhaustion is greater than the rate of its generation.

Solid, liquid, and gas phases are all important components of a soil. The solid phase is composed of mineral and organic matter distributed roughly 95 percent and 5 percent by weight, respectively, in an average soil but changed drastically in a reducing atmosphere. Muck soils and peats can contain from 6 percent to as little as 5 percent or less of mineral matter (2). Because of the great difference in density between mineral and organic matter, the organic matter (with only 5 percent of the weight) may occupy nearly 25 percent of the volume. The total volume occupied by the solid phase is frequently only about 50 percent of the total. The liquid and gas phases vary in proportion to rainfall, drainage, and plant usage. The composition of the soil air is generally the same as the atmosphere—80 percent nitrogen and 20 percent oxygen. When aeration is poor (that is, when the soil air has limited interaction with the atmosphere), the CO_2 content can increase from less than 0.1 percent to an extreme case of 20 percent (replacement of all of the oxygen). The increase in CO_2 is caused by the consumption of O_2 and the production of CO_2 by living soil organisms. A high water table can fill the pore spaces and remove both O_2 and CO_2. Some of the behavior of soil moisture has been discussed previously, as part of the water cycle. Water is the primary means of elemental transport in the soil system, providing nutrients and removing wastes. Generally the soil water can be considered a dilute solution (0.01 normal) of common salts of Na, K, Mg, Ca, Cl, NO_3, SO_4, HCO_3, etc. In saline soils, this concentration can increase ten-fold (to about 0.1 normal)—close to the wilting point for plants (3).

Every soil has a profile—a succession of layers (known as horizons) grading from a surface layer of organic material down into loose, weathered rock. As rain water filters through the organic material, it forms various organic acids, which remove soluble salts and material fine enough to form a suspension. The resulting leached layer is called the surface soil (or A horizon, or eluviated zone). The leached material is deposited in the layer below, commonly termed the B horizon of the subsoil. The organic layer, leached layer, and layer of deposition form the solum, the genetic soil, which is considered different from the C horizon (essentially weathered parent rock material) and the D horizon (unweathered rock that is not necessarily parent material). Not all soil horizons are present in any one soil profile, but every profile has some of them. The presence, thickness, and depth of the layers are a function of the soil-forming process.

Soil Formation

Although there is not agreement among various soils specialists concerning the complete list of factors responsible for the formation of soils, they generally agree on the following: nature of the parent material, climate, topography, time, and biologic activity. The following sections will discuss these factors in a broad way. (For a more detailed treatment, see Refs. 4, 5.)

Parent Material

The parent material provides an initial abundance of major and minor elements, which permits geochemical prospecting for ore bodies by soil sampling. Although one could argue that there should be as many different kinds of soils as there are kinds of parent rock, Table 4-1 indicates that such is not

TABLE 4-1
Chemical Elements Depleted and Concentrated in Soil

A. Elements depleted in soil relative to their crustal abundance

    Major   Ca, Mg, P, Na

    Minor   Cr, Co, Cu, F, Fe, Mn, Ni, V, Zn

    Nonesstential   Be, Cd, Hg, Rb, Sr, Ti

B. Elements concentrated in soil relative to their crustal abundance

    Major   C, N, K, S

    Minor   As, I, Mg, Se, Sn

    Nonessential   Sb, Ba, B, Br, Pb, Li, La, Zr

C. Elements neither depleted nor concentrated (Ratio = 0.80 – 1.20)

    Major   Cl, O

    Minor   Si

    Nonessential   Al, Ge, Th, Ag

Source: Soil composition (6, 7, 8 and 9). Crust composition (10). Major, minor, and nonessential elements are taken from Table 1-1.

the case, because on a worldwide average basis not all elements behave the same way. Some are depleted in the soil relative to their abundance in the crust, and some are concentrated. There is no simple explanation, because it depends on a complex interplay of the mode of occurrence of the various chemical elements, and of the stability of that mode to the weathering or soil-forming process. This will be discussed in the section on climate.

The concept of soil maturity refers to a thermodynamically stable end product of the weathering process—that is, only the most stable mineral species under the conditions of weathering are present. Recent work has emphasized that the proportions of Fe, Al, and Si oxides, hydroxides, and oxyhydroxides with the clays kaolinite and halloysite increase with soil maturity (mineralogical maturity) relative to primary quartz, feldspar, ferromagnesian minerals, and secondary chlorite, vermiculite, montmorillonite, and other clays (11,

12). As the secondary phases increase, the composition of the soil increases in SiO_2, Al_2O_3, and Fe_2O_3 (geochemical maturity). In addition, the concentration of SiO_2 will increase in the sand-sized fraction of the soil, Al_2O_3 will increase in the clay-sized fraction relative to the sand, and Fe_2O_3 will increase in the oxide fraction relative to the sand (13). In any event, weathering products should become more alike with time, and then it should be impossible to tell, on geochemical grounds, what the parent rocks were. Such has happened with laterites and bauxites produced from dunites, diabases, nepheline syenite, and contact metamorphic rocks in Africa; they have been weathered for more than a million years (12). The parent-rock effect is an inverse function of time, the greatest effect occurring early in the soil-formation process.

Time

If the parent-rock effect is indeed a function of time—that is, if all rocks tend to result in the same type of soil, given a sufficient duration of weathering—then time is the only independent variable in the soil-formation process. Humans are interested not only in processes that take millions of years, but also in processes that take place in their lifetimes. The amount of time needed for a particular soil to develop depends on where the process must start. With limestones, millions of years may pass before enough parent material has accumulated; generally the process of accumulation of soil takes orders of magnitude longer than for the differentiation of soil horizons. Development of a distinct leached layer can occur in tens of years, whereas a distinct B horizon (or depositional layer) may take hundreds or even thousands of years to develop. This is in contrast with the Moon, where it takes an average of 1 million years for 1.5 mm of soil to develop.

Slope

Very few soils are solely the response to vertical processes operating on level ground. Recent work (14) has focused on the combination and intensity of the processes that cause landform changes. Figure 4-2 and Table 4-2 illustrate 9 different land forms, the soil criteria that develop on them, and the processes that are operative. Slope clearly influences the moisture regime and the erosional pattern.

Biological Activity

Vertical variations in soil-profile development are common, being controlled not only by slope, time, and parent material, but also by organic matter. Indeed, the formation of soil profile and the development of vegetation are closely linked. As fresh rock is exposed, the first plants colonize the surface; with time, both vegetation and soil increase in complexity.

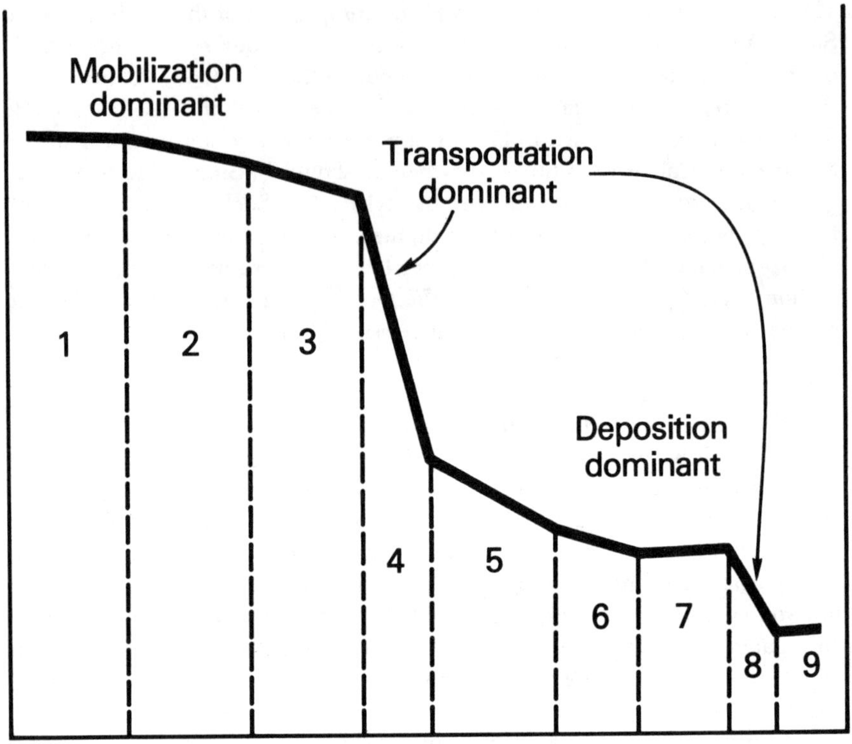

Mobilization dominant

Transportation dominant

Deposition dominant

1 2 3 4 5 6 7 8 9

See Table 4-2 for numbered columns.

FIGURE 4-2. Land-surface classification in relation to processes that cause changes in landforms. See Table 4-2. Source: Ref. 14.

Living organisms—plants, animals, insects, bacteria, fungi, and the like—are important chiefly to horizon differentiation and less so to the accumulation of soil parent materials. Gains in organic matter and nitrogen in the soil, gains or losses in plant nutrients, and changes in structure and porosity are among the shifts due to living organisms. Plants and animals may also mix horizons and thus retard their differentiation.

Plants largely determine the kinds and amounts of organic matter that go into a soil under natural conditions. They also govern the way in which it will be added, whether as leaves and twigs on the surface or as fibrous roots within the profile.

Some plants take their nitrogen from the air and add it to the soil as they die. Deep-rooted plants reverse leaching processes in part. The roots may take up calcium, potassium, phosphorus, and other nutrient elements from the C horizon or even from the deeper regolith, only to leave some part of those

TABLE 4-2
Land Form Units and Processes

Land Form Unit	Definition	Predominant Soil Criteria	Predominant Geomorphic Process
1.	Interfluve	Soil development in situ well drained	Vertical movement of soil water
2.	Leaching predominate	Gley soils, compaction, Mn concentration	Mechanical and chemical leaching by lateral soil water movement
3.	Soil creep	Substitution by upslope soil better drained than 2	Soil creep predominates
4.	Fall and rock slide	Little soil formation	Fall, slide, physical and chemical weathering
5.	Down slope transportation	Wash and creep dominate constrasting soils	Rapid mass movement and surface runoff
6.	Redeposition from upslope	Heterogenous mantle	Deposition from mass movement and surface runoff
7.	Redeposition from upvalley (flood plain)	Alluvium added, poorly drained	Subsurface soil and ground water movement, alluvian deposited
8.	Channel wall	Little soil formation	Slumping, lateral corrosion by stream, fall
9.	Stream channel bed	Absence of soil formation	Transportation downstream

Source: Ref. 14. See Figure 4-2.

nutrients in the solum when the plants die.

Horizons may be mixed by plants or animals. When trees tip over in a forest, the roots take up soil materials from several horizons. As the upturned roots decay, this soil material tumbles back down, mixing as it goes. Burrowing animals also mix horizons as they build their homes.

Bacteria and fungi live mainly on plant and animal residues. They break down complex compounds into simple forms as in the decay of organic matter. It has been suggested that the humus in soils is largely dead bodies of microorganisms; much of it seem to have about the same composition, even though it exists under widely different types of vegetation. Some micro-organisms fix nitrogen from the atmosphere and thus add it to the regolith in their bodies when they die (15).

The curves shown in Figure 4-3 illustrate that higher plants (which comprise more than 99 percent of the biosphere) begin their production of organic raw material with increasing temperature, reach maximum production at 25-30°C (80-90°F), and decrease production at higher temperatures, assuming sufficient air and water. In semiarid climates, grasses produced during the "wet" season suspend or limit their growth during the dry season. The growth curve for organic raw-material production in a semiarid climate would follow the same general shape as in a humid climate, but the quantities produced would be far less.

The destruction of organic matter and formation of humus is a function of the activity of microorganisms, which begins at about 5°C (40°F) — it is not by accident that we keep our refrigerators set at about this temperature (16). This means that within the range of 0 to 25°C (30 to 80°F) humus is produced and preserved; however, above 25°C (80°F) the activity of microorganisms exceeds the productivity of plants and no humus is accumulated, again assuming the presence of sufficient moisture and air.

A variation of the above is the situation of a high water table, such as a swamp, where microorganisms suffer from a deficiency of air and oxygen and the rate of microbial reactions may be reduced or completely eliminated. When O_2 is absent, new processes may occur that will be harmful to plants — e.g., production of CH_4. The activity of the organisms is never as great as the productivity of the plants, and large amounts of humus collect in a reducing environment. Histic soils result — for example, the underclay of coals is a fossil gley. (For a complete taxonomy of soils, see Ref. 17 and for a comparison of old and new terminology, see Ref. 5, ch. 7.)

Microorganisms also react to too little water, suffering much more than plants. This means that the activity of microorganisms in a semiarid environment is less than the productivity of plants; humus collects and dark, rich soils form. In arid climates, organic matter is not produced and the desert soils are light in color.

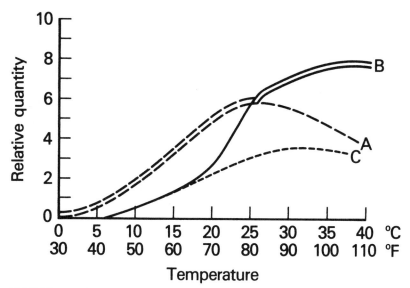

FIGURE 4-3. Organic matter production in humid climates (A) compared to that destroyed in aerated (B) and unaerated (C) humid climates. Source: Modified from Ref. 16.

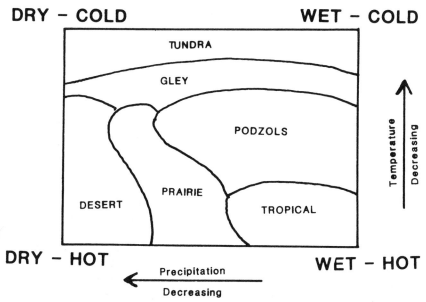

FIGURE 4-4. Soil-zone formation as a function of temperature and precipitation. Source: Modified from Ref. 18. See also Fig. 6-7.

Climate

Climate is generally considered to be the combined effect of temperature and precipitation and is a local physical manifestation of the worldwide water cycle. Because these factors are also the controls for vegetation development, it is impossible to separate out the effects caused by temperature and precipitation from the biological effects discussed previously. The general zones of soil that develop in response to great variations in temperature and precipitation are illustrated in Figure 4-4. The general characteristics of the soil types are quite distinctive (5, 17, 19). Tropical soils occur where an annual high temperature, combined with the lack of a severe winter, permit sustained bacterial action. This action destroys the plant litter and other humic materials as rapidly as they are produced. Little humic acid is produced. In its absence, oxides of iron and aluminum, which are insoluble, accumulate in the soil profile as red and yellow clays, nodules, and rock-like layers (limonite, bauxite, and laterite). Silica is soluble and, with Ca, Mg, Na, and K, is leached from the soil. Low soil fertility results because bases are not held and humus is lacking. Distinctive soil horizons are generally absent. If colloidal clay is absent, the soil tends to be firm and porous and will transmit water readily. With clays present, the soil is sticky and plastic, and deposition is slight.

A tropical soil can become dehydrated by change of climate or by some human-induced action that does not allow the water to percolate through the soil. Such action as stripping the vegetation increases the runoff to a value where runoff and evaporation are greater than downward percolation. The resulting irreversible dehydration of the iron oxyhydroxides leads to the concretions and hard nature characteristic of "dry" laterites.

Podzols (Spodosols), as indicated in Figure 4-4, develop in climates cold enough to inhibit bacterial action but with sufficient moisture for plant growth. The soil-moisture budget shows a surplus in the winter and spring with small deficiencies in the summer. Coniferous trees do not use large amounts of chemical bases, which are therefore not returned to the soil through litterfall (see Chapter 6 for a more detailed discussion). Humic acids strongly leach the upper soil of bases, colloids, iron, and aluminum, leaving only a gray silica-rich layer. The materials removed accumulate in a dark, dense B horizon. There is a twofold flow through the B horizon—when wet, the flow is downward from the humus-rich top layer carrying Ca, Mg, Na and K; when dry, the flow is upward, driven by surface evaporation such that Fe and Al oxyhydroxides accumulate near the top of zone B.

Prairie soils (Mollisols) occur in cool to hot climates with scant precipitation. The vegetation consists mostly of grasses. Zone A is rich in humus at the top and in bases below, with both downward and upward flows that carry the nutrient elements. Generally, evapotranspiration exceeds precipitation so that a water deficit prevails in at least one season and a water surplus at any time is negligible. The downward movement of water is not sufficient to leach out the bases, and grasses restore them to the soil. Calcium carbonate is brought upwards and precipitates in the B horizon during times of moisture deficiency, forming calcrete or caliche.

Humus occurs in progressively smaller amounts from the prairie soils toward the desert soils (Aridosols). There is meager litter produced from the sparse vegetation in these soils and the high temperatures generally cause rapid decomposition.

Under cold conditions, the organic matter can accumulate faster than it is oxidized, and peat tends to form. Even though precipitation is low, evapotranspiration is less, and soil-moisture storage remains adequate throughout the year. If the area is poorly drained, bogs can accumulate on a gley horizon—a compact, sticky, structureless clay. The blue-gray color of the clay comes from reduced iron in contrast to the yellow, brown, or red of oxidized iron. For still colder climates such as tundras, there is little vegetation. That which does fall as litter tends to accumulate because it is too cold for humification. The balance beween plant, litter, and soil organic matter is discussed in more detail in Chapter 6 (for example, Table 6-3 and Figure 6-9).

The generalized world distribution of the main soil types is illustrated in Figure 4-5. Knowledge of the soil regimes and their associated soil-moisture budget allows evaluation of potential for food production. The map indicates that areas of optimal precipitation, temperature, and soil are limited. The importance of soil in geochemical exploration justifies a much longer treatment (5).

Soil Composition

It is evident from the preceding sections that any global discussion of soil composition is fraught with conceptual difficulties. Nevertheless, there have been several major efforts io accumulate and analyze a sufficient number of samples to show global trends and patterns, if not to provide the quantification necessary for the calculation of global cycles. Ranges of soil compositions are given in Figure 4-6, which also shows the distinct differences in the mean compositions of soils in the eastern and western United States. These differences were previously indicated in Figure 4-5. The eastern United States is characterized by podzols and laterites, whereas calcification is the process affecting the bulk of the western U.S. soils. It is of interest to compare the range of soil compositions with the composition of the crust (Figure 1-2) and of sea water (Figure 5-2).

Similarity of soil enrichment can, in many cases, be ascribed to similarity of chemical solubility in the soil water. Mineral matter present in soil is a

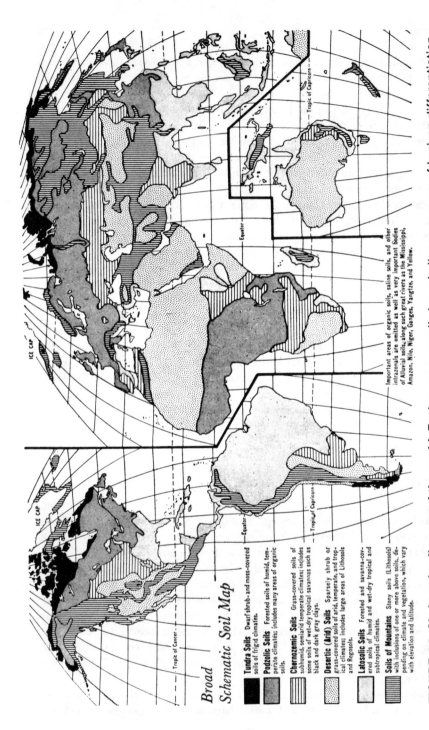

FIGURE 4-5. Six broad soil zones of the world. Each zone generally has similar processes of horizon differentiation prevailing over it. Many kinds of soils are present in every zone. Source: Ref. 15.

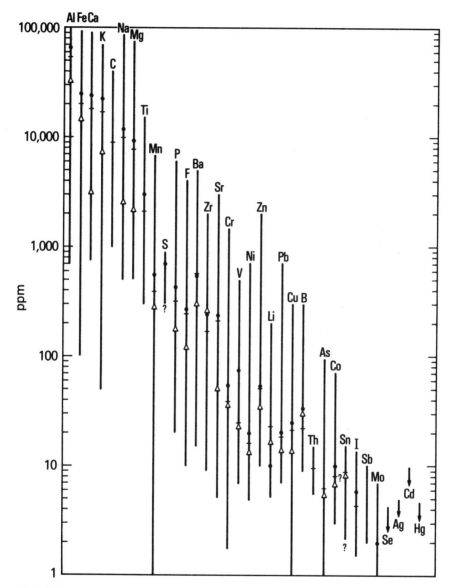

FIGURE 4-6. Ranges of measured abundance of elements in soils. Dots (Ref. 5) give overall average. Triangles indicate eastern U.S. averages and a short dash indicates western U.S. averages (Ref. 17). Sources: Information from Refs. 4, 6, 7, 8, 9, 18, 20, 21.

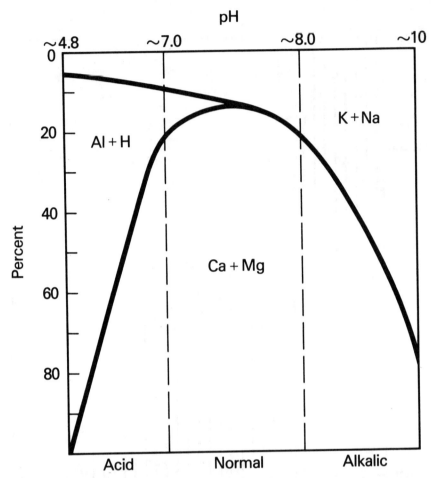

FIGURE 4-7. Variation in major cation composition of soils with changing acidity. Source: Ref. 22.

function of its stability to solution processes. If the minerals are completely unstable (highly soluble), such as nitrates, most of the halides, and some sulfates, they will be present only under the most extreme conditions—saline soils. Minerals of low and intermediate solubility provide the major elements available in the soil. In normal arable soils, 90 percent of such elements consist of Ca, Mg, Al, and H with the remainder being Na, K, and sometimes NH_4^+. The proportions of these elements change drastically as the acidity (pH) varies, as indicated in Figure 4-7.

Organic matter present in soil can be classified into two groups:

a. Compounds of a nonspecific nature related to various classes of organic chemistry, such as decomposition products of organic plant and animal residues and elements of the plasma of microorganisms (proteins, carbohydrates, organic acids, fats, waxes, resins, etc.). These compounds form 10–15 percent of the total soil organic matter.
b. Humic substances that are specific, high-molecular weight, complex compounds, and appear to fall into two groups: humic acids and fulvic acids. The fulvic acids are the simplest, being soluble in water and having a slight yellow color (fulvus = yellow). They assist in the mobilization of Fe and Al, whereas humic acids leach fewer metals from the soil profile. Fulvic acids are twice as abundant as humic acids in podzols, and they are about equal in the prairie soils.

Mode of Occurrence

Trace elements in soils follow the same variations in mode of occurrence as in stream sediments. Table 4-3 indicates the four major modes: parent minerals not yet altered, major and minor; a mineral formed during the soil-forming process; and an element not structurally bound but absorbed on a surface. The initial distribution of the trace element into the parent mineral is not part of this study. It would be of value to know how each element is distributed among these four different modes, and in how many sites in each of these modes. As mentioned in the discussion of stream sediments, such information is just now becoming available. It does seem to indicate that a given element can be present in all modes, although it might prefer a single one. For example, extraction-rate analysis for two forest soils has indicated that Na, K, Ca, and Fe are held in four separate sites in the different modes (24).

Mobility of Elements

Living vegetation, bacteria, algae, fungi, and other microorganisms all exert a profound influence on the distribution of chemical elements. Roots of plants provide H^+ by osmosis while taking up the nutrients Ca, Mg, and K. Uptake of inorganic matter is distributed through the leaves, stems, etc., of the plant, which, in turn, fall to the ground and decay. Rain water leaches the soluble constituents. Part can be taken up again by the plant and another part can be absorbed in the B horizon with Fe, Mn, and Al. The less-soluble constituents released by plant decay tend to remain in the humus layer. This effect is cumulative, and the biogenic enrichment that results is known as the

TABLE 4-3
Mode of Occurrence of Trace Elements in Soil and Stream Sediments

Mode	Constituent Element in Trace Mineral	Trace Constituent in Major Mineral	Trace Constituent in Mineral Formed During Weathering	Absorbed on Surface
Example	Cu in malachite Pb in anglesite Au	Zn in magnetite Pb in K-feldspar Cu in biotite	Zn in montmorillonite Co, Cu in Fe-Mn oxides Hg in organics	Fe-Mn oxide Colloidal clay Colloidal organic Clay exchange layer
Parameters affecting mobility	Simple solubility and solution chemistry	Properties of host: will be sorted and partitioned by grain size, density, and chemical behavior and slowly released	Tend to be more accessible to surrounding solutions	Ion exchange equilibria

Source: Ref. 23.

Goldschmidt-reaction principle. Microorganisms also play a major part in the cycles of carbon, nitrogen, phosphorus, sulfur, and iron (25) and have strong effects on elements possessing variable oxidation states by reducing them. Those elements affected are not only Fe and Mn, but also As, Mo, Se, etc., which occur as mobile anionic complexes and immobile cations. The entire process is known as the biogeochemical cycle.

Adsorption. Generally the inorganic constituents can be separated with regard to solubility and adsorption behavior. Adsorption behavior is primarily a function of the amount of surface area. Coarse minerals with small surface areas are not important quantitatively in the adsorption process. Materials with large surface areas must have low solubility because there is usually a very rapid disappearance of very small particles of a highly soluble salt (3). Generally the amount of surface area that limits the importance of the process is 1 m^2/g, corresponding to a particle diameter of 2 μm. Clay and hydroxides of Fe and Al often have surface areas of several tens of square meters per gram. Clay minerals adsorb cations by virtue of the charges that exist both at the edges of the crystal flakes and within the structural layers of the minerals, with adsorption increasing as the size of the flakes decreases (because more edges are present). Some anions can be adsorbed by the clays through replacement of (OH) groups. Ions appear to be held by organic matter as carboxylic acid radicals, attached to the extremely large complex organic molecules of the humus. Exchange capacity of humus increases with increasing pH (more basic) and, in general, far exceeds that of clay minerals. Humic substances are soluble in acid or alkaline media, but lose their mobility in neutral or weakly alkaline media. Finally, in natural solutions the colloidal silica, hydrous manganese oxide, and humic colloids are negatively charged and can adsorb cations (see Figure 3-11). Colloidal alumina is positively charged and can adsorb such anions as PO_4^{3-}, VO_4^{3-}, AsO_4^{3-}, and SO_4^{2-}. Colloidal iron oxides may be either positive or negative, and usually adsorb anions. A large cation-exchange capacity is generally considered a beneficial characteristic of the soil system: retention of fertilizer cations will decrease leaching losses, and polluting cations can be retained in the soil instead of being removed by ground water. Although most of the effort in analyzing adsorption has focused on the behavior of clay minerals, most adsorption is connected with the organic and oxhydroxide phases. The variation of these phases with soil type has been previously discussed.

Solubility. Solubility varies greatly with oxidizing environment, acidity (pH), organic acid content, temperature, and many other factors. Stable minerals generally have small surface areas, whereas minerals formed during the weathering process have large surface areas. The breakdown of the rock minerals in the soil thus liberates many of the elements needed by plants. Predicting the persistence of a mineral has occupied many geologists over

TABLE 4-4
Relative Mobility of Elements with Variation in Acidity and Oxidizing Conditions

Mobility	Environmental Conditions		
	Oxidizing (Acid)	Oxidizing (Neutral-Alkaline)	Reducing
Very high	Cl,I,Br	Cl,I,Br	Cl,I,Br
	S,B	S,B	
		Mo,V,U,Se	
	Rn,He	Rn,He	Rn,He
High	Mo,V,U,Se		
	Ca,Na,Mg,Li, Sr,Ba,F	Ca,Na,Mg,Li, Sr,Ba,F	Ca,Na,Mg,Li, Sr,Ba,F
	Zn		
	Cu,Co,Ni,Hg,Ag,Au		
Low	As,Cd	As,Cd	
	Si,P,K	Si,P,K	Si,P,K
	Pb,Be,Bi,Sb,Ge,Tl	Pb,Be,Bi,Sb,Ge,Tl	
	Fe,Mn	Fe,Mn	Fe,Mn
Very low to Immobile			S,B
			Mo,V,U,Se
		Zn	Zn
		Cu,Co,Ni,Hg,Ag,Au	Cu,Co,Ni,Hg,Ag,Au
			As,Cd
			Pb,Be,Bi,Sb,Ge,Tl
	Al,Ti,Sn,W,Nb,Ta Pt,Cr,Zr,Th Rare Earths.	Al,Ti,Sn,W,Nb,Ta, Pt,Cr,Zr,Th Rare Earths	Al,Ti,Sn,W,Nb,Ta, Pt,Cr,Zr,Th Rare Earths.

Source: Refs. 5, 27, 28, 31.

the years (26), and the oxidizing environment (Eh) and acidity (pH) of some natural environments have been used to indicate the relative mobilities of elements in water as the environment varies (5, 27–31). Table 4-4 illustrates the grouping of geochemical elements and their mobility under the varying conditions.

Recent work (32) has shown two distinct soil compartments for the movement of materials. The first, from the top of the organic layer through the bottom of the B horizon or depositional layer, is controlled by organic acids that lower the pH of the soil solution and depress the bicarbonate concentration. Presence of Al and Fe in the fulvic acid spectra illustrates the role that the organic components take in the mobility of these elements. There is a 60 to 70 percent decrease of the fulvic acids across the depositional layer. In the lower compartment, bicarbonate dominates the soil solution and the pH is higher. Small grains of chlorite, vermiculite, and feldspar are mobilized in the zone of deposition, and transported to depth. Thus, the B horizon of a podzol is a zone of deposition for suspended organics and their complexed metals, and a zone of leaching for the silicates.

Relative effects of organic acids, bicarbonate, and anthropogenic mineral acids introduced through acid rain have recently been studied (33). Carbonic acid leaching predominates in tropical and warm-temperature soil profiles and in deeper horizons of cooler soils (see above); organic acid leaching occurs in cool-temperature, arctic, and alpine acid soils; and H_2SO_4 and HNO_3 leaching predominates in the mountains of New England and may be the dominant process in regions of the northeast United States affected by acid rain. The relative efficiency of the mineral acids compared to organic acids in the leaching process is not yet known.

Mass wastage. The last major source of mobility of elements to be discussed is mass wastage—the destruction and removal of the soil. Erosion originates from two basic sources, cropland and noncropland. From the standpoint of agricultural policy the problem of erosion is divisible into controllable and uncontrollable erosion (34). Erosion from cropland can be calculated by using the universal soil loss equation (35) discussed previously. Of the variables in the equation, only the crop-management tillage factor is amenable to control (36).

Control of erosion from croplands may have only a marginal effect on sediment loads in streams. More than 70 percent of the national total annual sediment load is contributed by noncropland sources. In some areas the proportion is so high that reducing cropland erosion has no effect on total load. In other areas, such as the Missouri River basin, predicted sediment load can increase to 5 times the historic load because of changes in land usage. If any costs of sediment control are to be apportioned (1) it is critical to define and measure all variables and set a historical and geological load value.

The problem of measuring the geologic load has been discussed, with the historic load generally 3 times the natural or geologic load. Rates of soil erosion for the Black Sea drainage area were measured by calculating the rates of sedimentation in sediment cores (37). No apparent rate change occurred before about 200 A.D. (going back to 4100 B.C.), with major variations apparently due to agricultural activities occurring after 1000 A.D. The variations average 3 times the apparent natural or geologic load, but can range up to 6 times that background for certain periods of apparently high flooding.

The previous chapter and this one have indicated that the bulk of the sediment generated by weathering does not reach the ocean, but that the bulk of the material dissolved in stream water does. The ocean reservoir is thus obviously the next subject for discussion in our cycle.

References Cited

1. General Accounting Office, 1977, To Protect Tomorrow's Food Supply, Soil Conservation Needs Priority Attention; Report CED-77-30, 59 p.

2. Jackson, M. L., 1964, Chemical Composition of Soils; p. 71-141 in Chemistry of the Soil, 2nd ed., F. E. Bear, ed., Reinhold, New York, 515 p.

3. Bolt, G. H., and M.G.M. Bruggenwert, 1976, Composition of the Soil; p. 1-12 in Soil Chemistry A. Basic Elements, G. H. Bolt and M.G.M. Bruggenwert, eds., Elsevier, Amsterdam, 281 p.

4. Hunt, C. B., 1972, Geology of Soils; W. H. Freeman & Co., San Francisco, 344 p.

5. Rose, A. W., H. E. Hawkes, and J. S. Webb, 1979, Geochemistry in Mineral Exploration; 2nd ed., Academic Press, London, 657 p.

6. Shacklette, H. T., J. Hamilton, J. Boerngen, and J. Bowles, 1971, Elemental Compositions of Surficial Materials in the Conterminous United States; U.S. Geological Survey Professional Paper 574-D, Washington, D.C., 71 p.

7. Vinogradov, A. P., 1959, The Geochemistry of Rare and Dispersed Chemical Elements in Soils; Consultants Bureau, Inc., New York, 191 p.

8. Bowen, H.J.M., 1966, Trace Elements in Biochemistry; Academic Press, London, 241 p.

9. Bowen, H.J.M., 1979, Environmental Chemistry of the Elements, Academic Press, New York, 334 p.

10. Beus, A. A., and S. V. Grigorian, 1975, Geochemical Exploration Methods for Mineral Deposits; Applied Publications Ltd., Wilmette, Ill., 287 p.

11. Chesworth, W., 1973, The Residua System of Chemical Weathering: A Model for the Chemical Breakdown of Silicate Rocks at the Surface of the Earth; Journal of Soil Science, v. 24, no. 1, p. 69-81.

12. Chesworth, W., 1973, The Parent Rock Effect in the Genesis of Soil; Geoderma, v. 10, p. 215-225.

13. Wakatsuki, T., H. Furukawa, and K. Kyuma, 1977, Geochemical Study of

the Redistribution of Elements in Soils — I. Evalution of Degree of Weathering of Transported Soil Materials by Distribution of Major Elements Among the Particle Size Fractions and Soil Extract; Geochimica et Cosmochimica Acta, v. 41, p. 891-902.

14. Conacher, A. J., and J. B. Dalrymple, 1977, The Nine-Unit Land Surface Model: An Approach to Pedogeomorphic Research; Geoderma, v. 18, p. 1-154.

15. Simonson, R. W., 1957, What Soils Are; p. 17-31 *in* Yearbook of Agriculture 1957, Soil; U.S. Department of Agriculture, Washington, D.C., 784 p.

16. Senstius, M. W., 1958, Climax Forms of Rock-Weathering; American Scientist, v. 46, p. 355-367.

17. Soil Survey Staff, 1975, Soil Taxonomy; U.S. Department of Agriculture, Handbook 436, Washington, D.C., 754 p.

18. Brooks, R. R., 1972, Geobotany and Biogeochemistry in Mineral Exploration; Harper and Row, New York, 290 p.

19. Strahler, A. N., and A. H. Strahler, 1973, Environmental Geoscience: Interaction Between Natural Systems and Man; Hamilton, Santa Barbara, Calif., 511 p.

20. Connor, J., and H. T. Shacklette, 1975, Background Geochemistry of Some Rocks, Soils, Plants, and Vegetables in the Conterminous United States; U.S. Geological Survey Professional Paper 574-F, Washington, D.C., 168 p.

21. Mitchell, R. T., 1964, Trace Elements in Soils; p. 320-368 *in* Chemistry of the Soil, 2nd Ed., F. E. Bear, ed., Reinhold, New York, 515 p.

22. Bolt, G. H., M.G.M. Bruggenwert, and A. Kamphorst, 1976, Absorption of Cations by Soil; p. 54-90 *in* Soil Chemistry A. Basic Elements, G. H. Bolt and M.G.M. Bruggenwert, eds., Elsevier, Amsterdam, 281 p.

23. Rose, A. W., 1975, The Mode of Occurrence of Trace Elements in Soils and Stream Sediments Applied to Geochemical Exploration; p. 691-705 *in* Geochemical Explorations 1974, I. L. Elliott and W. K. Fletcher, eds., Association Exploration Geochemistry Special Publication No. 1, Elsevier, New York, 720 p.

24. Thompson, G. R., M. Behan, J. Mandzak, and C. Bowen, 1977, On the Evaluation of Nutrient Pools of Forest Soils; Clays and Clay Minerals, v. 25, p. 411-416.

25. Alexander, M., 1977, Introduction to Soil Microbiology; 2nd ed., J. Wiley and Sons, New York, 467 p.

26. Curtis, C. D., 1976, Stability of Minerals in Surface Weathering Reactions: A General Thermochemical Approach; Earth Surface Processes, v. 1, p. 63-70.

27. Hawkes, H. E., and J. S. Webb, 1962, Geochemistry in Mineral Exploration; Harper and Row, New York, 415 p.

28. Andrews-Jones, D. A., 1968, The Application of Geochemical Techniques to Mineral Exploration; Mineral Industries Bulletin, Colorado School of Mines, v. 11, n. 6, p. 1-31.

29. Swaine, D. J., and R. L. Mitchell, 1960, Trace-Element Distribution in Soil Profiles; Journal of Soil Science, v. 11, no. 2, p. 347-368.

30. Mitchell, R. L., 1972, Trace Elements in Soils and Factors That Affect Their Availability; Geological Society of America Special Paper 140.

31. Perel'man, A. I., 1967, Geochemistry of Epigenesis; trans. by N. N. Kohanowski, Plenum Press, New York, 266 p.

32. Ugolini, F. C., H. Dawson, and J. Zachara, 1977, Direct Evidence of Particle Migration in the Soil Solution of a Podzol; Science, v. 198, p. 603-605.

33. Cronan, C. S., W. Reiners, R. Reynolds, Jr., and G. Lang, 1978, Forest Floor Leaching: Contributions from Mineral, Organic, and Carbonic Acids in New Hampshire Subalpine Forests; Science, v. 200, p. 309-311.

34. Wade, J. C., and E. O. Healy, 1978, Measurement of Sediment Control Impacts on Agriculture; Water Resources Research, v. 14, n. 1, p. 1-8.

35. Wischmeier, W. H., and D. D. Smith, 1965, Predicting Rainfall-Erosion Losses from Cropland East of the Rocky Mountains; U.S. Department of Agriculture, Handbook 282, 47 p.

36. Agricultural Research Service, 1975, Present and Prospective Technology for Predicting Sediment Yield and Sources; U.S. Department of Agriculture, ARS Bulletin ARS-S-40, 285 p.

37. Degens, E. T., A. Paluska, and E. Eriksson, 1976, Rates of Soil Erosion; p. 185-191 *in* Nitrogen, Phosphorus and Sulfur—Global Cycles. SCOPE Report 7, Ecology Bulletin, Stockholm, v. 22, 192 p.

5
Ocean Reservoir

An ocean is forever asking questions.
—E. A. Robinson

Introduction

The oceans are the major reservoir of water on the Earth's surface (Figure 5-1). The fluxes of material into and out of the ocean, and the sources of those fluxes, have been described. We will now consider the chemical composition of sea water, the behavior of the major nutrient elements, the suspended sediments, and the sediments on the ocean bottom. The ranges of chemical composition of average sea water are shown in Figure 5-2, which gives published values for sea water with a salinity of 35‰. Water in estuaries shows a complex mixing that depends on the relative strengths of the rivers, with resulting mixing of salinity values. Regions of high evaporation also show an increase in salinity as the dissolved salts are left behind. Both processes would be expected to produce salinity variations in the oceans. It is therefore startling to observe that the salinity is relatively constant from ocean to ocean and at different depths; values are as low as 1‰ at fresh-water inputs and as high as 37‰ at evaporation centers. Other variables that influence geochemical processes in marine systems are temperature (-2 to $27°C$), pressure (1 to 1000 atm), pH (7.3 to 8.4), and redox potential (-0.3 to $+0.8$ volts).

Chemical Composition

Elements present at concentrations of more than 1000 $\mu g/l$ (ppb) contribute to the salinity in a measurable way and are described as major elements. Dissolved Si shows a wide variation but is generally excluded from the list of major elements. Elements Na, Mg, K, S, Br, and Cl occur in constant pro-

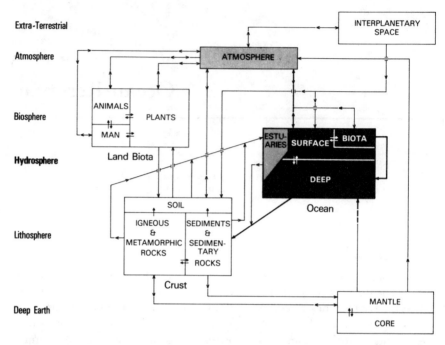

FIGURE 5-1. The ocean reservoir in the geochemical cycle. The ocean reservoir is in black. Shaded reservoirs indicate those with major interactions with the ocean reservoir. Important fluxes are indicated by heavy lines.

portion to each other (within limits of analytical uncertainty), whereas Ca shows enrichment with depth. The constant ratio indicates that the residence times are much longer than the mixing times of the oceans, and such elements are called conservative. The only chemical compound that is present at saturation amounts is $CaCO_3$, and the Ca content will vary as the carbonate is precipitated and dissolved.

As mentioned previously, it is important to determine the different chemical species because they govern the chemical behavior of the elements in sea water. Copper as a free ion is more toxic than copper in organic complexes, whereas the reverse is true for mercury. Nitrogen present as NH_3 is more toxic to fish than as NH_4^+. As expected, ions of the major elements have been extensively studied, although organic acids, ammonia, and silicate species are still largely unknown; furthermore, greater attention must be given to the chloride ion, which is the most common ion in sea water. The many analytical

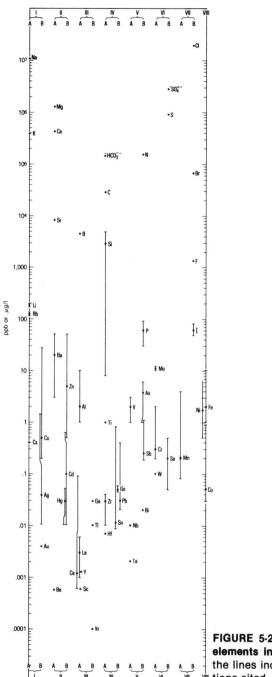

FIGURE 5-2. The concentration of selected elements in average sea water. Length of the lines indicates the range of concentrations cited. Sources: Refs. 1, 2, 3, 4, 5, 6.

problems and many assumptions make measurement of chemical species an area of controversy.

Minor or trace elements are present in amounts so small as to push the limits of analytical capability (Figure 5-2), with the range of compositions covering 12 orders of magnitude — 1000 times the range found in crustal rocks. As Turekian stated (7),

> contemporary geochemists . . . keep finding lower and lower concentrations of potentially hazardous (or beneficial) metals in the sea as their methods improve even as the rate of worldwide use of these metals is increasing. Indeed, one has the feeling that the whole field of trace metal marine geochemistry would have been a completely dull one over the past fifty years if it weren't for analytical errors.

He went on to stress that not only are most trace metals present at extremely low concentrations in the ocean but that they also have "rather unspectacular variations" in their concentrations. This emphasizes the importance of adsorption by particulate matter as a scouring mechanism for the removal of such elements.

Chemical Speciation in Sea Water

The chemical form of a particular element will help control the reactions, be they adsorption-desorption or precipitation-dissolution. There are generally two ways to determine speciation (1), that is, the ionic complexes at which a particular element occurs. The first uses analytical measurements after various physical and chemical separations. The second is an operational definition derived by measuring the response to perturbations such as pH change or oxidation. The major cations and anions are mainly free cations or anionic complexes, as shown in Table 5-1.

For example, 98 percent of sodium occurs as Na^+ with the other 2 percent primarily as $NaSO_4^-$. Similarly, about 90 percent of magnesium occurs as Mg^{2+} with the remainder also a sulfate and minor, but measurable, amounts as bicarbonate. Calcium occurs about 90 percent as the free ion species Ca^{2+}, about 10 percent as sulfate, and also as measurable amounts of bicarbonate and carbonate. About 50 percent of the carbonate species present is $MgCO_3$ followed by $CaCO_3$, Na_2CO_3, and free CO_3^{2-}. For SO_4^{2-} and HCO_3^-, the free anion complex is quantitatively the most abundant species.

While there is generalized agreement about major ion speciation, the range for trace elements varies greatly. Several elements are indicated in Table 5-2. For iron, the range for the electrically neutral species, $Fe(OH)_3$, is 0 to 95 percent; for the positively charged species it varies from 5 to 84 percent, and for

TABLE 5-1
Major Ion Speciation in Sea Water

		Anion Complex		
		Sulfate	Bicarbonate	Carbonate
Cation	–	SO_4^{2-}:39–54	HCO_3^-:63–81	CO_3^{2-}:8–9
Sodium	Na^+:98	21–40 / 2	11–20 / –	3–16 / –
Magnesium	Mg^{+2}:87–92	15–22 / 8–12	6–14 / 0.1–1.0	44–50 / –
Calcium	Ca^{+2}:85–92	3–5 / 8–13	1.5–3 / 0.1–0.5	21–38 / 0.1–0.9
Potassium	K^+:98–99	– / 1–2		

Source: Ref. 1, 3, 8, 9, 10.

Key: 8–12% of Mg occurs as $MgSO_4$
15–22% of (SO_4) occurs as $MgSO_4$

the negatively charged species from 0 to 40 percent. Predicting the adsorption possibilities of iron is dependent on the species present—wildly varying predictions could result from the choices taken. On the other hand, the values for cadmium are consistently 40 percent neutral, and 30 percent each for the positively and negatively charged species. Cadmium should then be expected everywhere! Not the least of the analytical problems is that particulate matter can pass through the filters, as indicated in Figure 5-3. Thus, a particular element could be included as particulate matter and not be distinguished from the same element in true solution.

Organic Matter

Of the chemical species in solution, organic complexes are the least understood and potentially the most important. The ocean may be visualized (11) as a slightly alkaline, very dilute solution of organic chemicals, reactive transition-metal ions, dissolved oxygen, and possibly active organic and inorganic solid phases whose surfaces are exposed to sunlight. The organic

TABLE 5-2
Range of Estimates of Speciation of Selected Trace Elements in Sea Water

Cadmium:	$CdCl^+$ (29-40%), $CdCl_2$ (38-50%), $CdCl_3^-$ (6-28%)
Chromium:	$Cr(OH)_2^+$ (85%), CrO_2^- (14%), CrO_4^{2-} (94%)
Copper:	$Cu(OH)_2$ (0-83%), $CuOHCl$ (0-65%), $CuCO_3$ (11-22%) $CuCl^+$ (6%), $CuOH$ (4%)
Iron:	$Fe(OH)^+$ (84%), $Fe(OH)_2^+$ (5-60%), $Fe(OH)_3$ (0-95%), $Fe(OH)_4^-$ (40%)
Lead:	$PbCl^+$ (11-19%), $PbCl_2$ (5-42%), $PbCO_3$ (0-76%)
Zinc:	Zn^{2+} (16-22%), $ZnCl^+$ (8-44%), $ZnCl_2$ (0-15%), $Zn(OH)_2$ (0-50%)

Source: Refs. 1, 8.

matter has several sources. The first is that produced by the photosynthesis process and released by the resulting organism into the water either as excretions or as dead cells. The second source is terrigenous humic substances carried into the ocean by rivers. Because of the flocculation caused by increasing pH from stream to ocean, the humic material has been suggested as an effective scavenging agent for metals. Another source comes from bottom-dwelling organisms, which excrete chemicals that combine with amino acids, sugars, and other compounds to form *gelbstoffe* (complex colored organics that are relatively resistant to microbial breakdown). The organic particulate form that results may be utilized by detrital feeders (11).

Removal of the organics from the ocean can occur at the ocean-atmosphere interface, when organics trapped in aerosols formed from wave action are physically removed. Photochemical degradation will break down aromatics easily with production of fatty acids. The fatty acids can readily be adsorbed

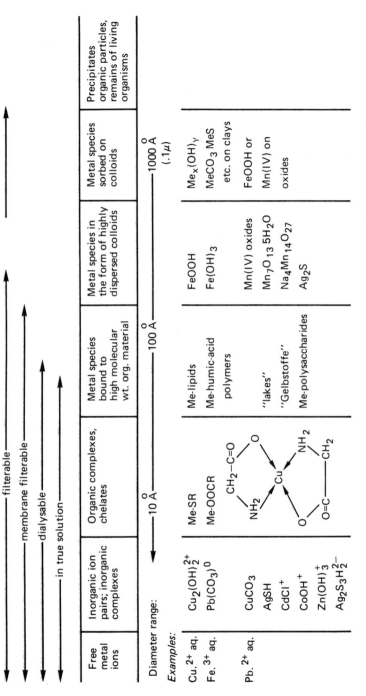

FIGURE 5-3. Forms of occurrence of metal species. Source: Ref. 10. Copyright 1975 by Academic Press, Inc. (London) Ltd. Published with permission.

onto clays and other materials, where it could interfere with the calcium-equilibrium adjustment.

Analysis of organic carbon in sea water is hindered by the large quantities of inorganic salts, which dwarf the organic quantities (12). The analytical process generally used removes all volatile organic compounds in the preparation, so that all information about them is lost. They are generally assumed to be less than 10 percent of the organic carbon.

Generally, the surface concentration of organic carbon in open water rarely exceeds 1500 µg/l, and below 500 m depth the concentrations are usually between 500 and 800 µg/l. Compare this with the inorganic load of 28,000 µg C/l. Practically all deep-water organic carbon is dissolved, with the particulate organic fraction amounting to only 10 µg C/l (3). Surface waters average 52 µg C/l (13), with nearshore waters having higher concentrations than deep-ocean surface waters. The distinction between particulate and dissolved carbon is based on the amount that passes through a 0.45 millipore filter (4500Å), which is off the scale on the right of Figure 5.3! Organic analyses of sea water for specific compounds or classes of compounds show that total carbohydrates are dominant, with unstable compounds such as amino or fatty acids amounting to only a small fraction of the organic carbon content (5). The material is comprised of compounds having a wide range of molecular weights (100–10,000) and with hydroxyl- and carboxyl-active groups. The term *humic material* is often used to describe this major fraction by uncertain analogy with the fulvic and humic acids in fresh water. The term *fulvic* would perhaps be more appropriate, because it is water soluble.

The chemical compositions of terrestrial (vascular plants) and marine (algae) productive systems differ greatly. Lignin and cellulose are the most continuously cycled organic materials because they are resistant to attack by microorganisms, but they are essentially absent from marine biota, which are composed of as much as 80 percent protein and simple carbohydrates with the soluble macromolecular fraction consisting of only 10 percent of the total organic material (14). The terms *humic* and *fulvic* therefore appear to resemble the terrestrial compounds only by analogy and should not be used for the marine environment. There is little or no direct evidence of organic complexing of metals in sea water. Stability-constant calculations indicate that the inorganic complexing discussed in the previous section is predominant unless unreasonably high organic concentrations are assumed. However, if the reality of organic complexes can be established experimentally, the calculations will undoubtedly be redone (14). For aerosols, the level of organic concentration is increased to several orders of magnitude beyond that considered by the equilibrium calculations, and the surface reaction might be the most fruitful place to search for these organic complexes.

Nutrient Elements and Chemical Reactions

The patterns of distribution of nutrient elements show several features in Figure 5-4. First, there is surface depletion of P, N, and Si. Second, there is an oxygen minimum. Third, there are relatively constant deep-sea concentrations, with the value for the Pacific Ocean higher than that for the Atlantic.

The elements N, P, and Si are depleted in surface water as a consequence of their utilization in biota production. With death and subsequent sinking of detrital organic matter, several processes take place that release these elements to sea water. First, the silica and carbonate of skeletal material tend to dissolve with depth, thus returning material to the ocean. Second, oxidation of the organic matter takes place approximately as:

$$(CH_2O)_{106}(NH_3)_{16}H_3PO_4 + 138\ O_2 = 106\ CO_2 + 122\ H_2O + 16\ HNO_3 + H_3PO_4$$

The oxidation of organic matter thus results in the production of water, carbon dioxide, and other soluble materials.

The close relationship between increases in nutrient content and depletion of free oxygen is obvious (3). Another relationship is the increase of hydrogen atoms to chemically balance the alkalinity of the ocean with the sinking of organic matter into deeper water. This is caused by dissociation of HNO_3, and H_3PO_4. Oceanic productivity depends on the rate and location of return of these nutrients to surface waters (15).

Trace elements do not seem to be enriched relative to phosphorus in plankton as compared to sea water (8), with the exception of Cu (a key element in photosynthesis). Many trace metals such as Sr, Ba, Ra, Cu, Ni, and Cd appear to be linked with nutrient elements, and dampen the distribution just presented. Sr exists as $SrSO_4$ shells and skeletons as well, and follows the pattern for Si for the same reasons—dissolution rather than oxidation. The linkage with nutrient elements is independent of the point of entry of the element and of the place of deposition, and therefore disguises any analysis of origin or destination. For some heavy elements, the removal agent is by particle setting rather than by biological concentration and removal (8). The nutrient-element pattern is not followed. Scavenging continues with depth, depending only on the presence of particles.

Oxygen abundance affects both pH and oxidation potential. Concentrations of O in surface zones can exceed saturation levels because of photosynthetic production. The reaction of O with H to form water decreases the acidity and raises the pH as high as 8.4. A pH value of 9 has been reported from one estuary having exceptionally high biological activity. The oxidation potential in the ocean is assumed to be controlled by this reaction. Deviations

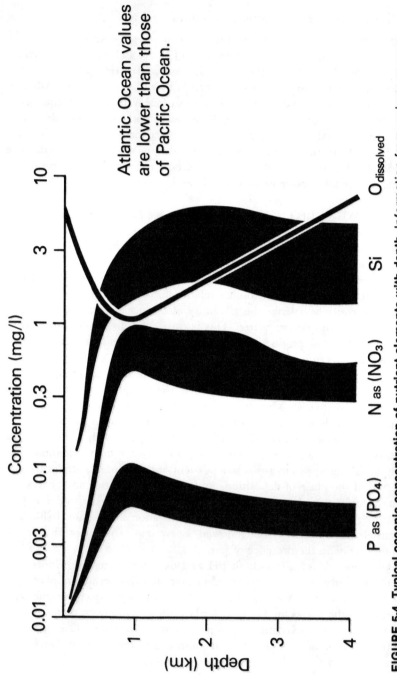

FIGURE 5-4. Typical oceanic concentration of nutrient elements with depth. Information from various sources.

do occur in some severely oxygen-depleted intermediate depth zones, in anaerobic conditions in some static basins such as the Black Sea, and at or below the ocean-sediment surface. The effect is that element species tend to be present in their most oxidized state—e.g. $(CrO_4)^{2-}$ instead of $Cr(OH)_2^+$. The Cr would tend to be present as Cr^{3+} rather than Cr^{6+} in the reducing environments mentioned.

There is a specific sequence of oxygen sources in sea water for the decomposition of organic material by microorganisms (16). After free oxygen is completely used, nitrate $(NO_3)^-$ is reduced to nitrite $(NO_2)^-$, a reaction that is used by some bacteria as an energy source. After about 60 percent of the nitrate is reduced, the nitrite begins to be reduced to free nitrogen (N_2). There is not a high concentration of the N-bearing complexes, so the process is balanced with the sulfur sequence.

$$\text{Organic} + (SO_4)^{2-} = CO_2 + 2H_2O + S^{2-}$$

This introduces new possibilities for removal of elements by formation of insoluble sulfides, and certainly a change in speciation for such elements as Fe, Mn, and I. Such a major redox boundary is unusual in the water column, but can occur widely in sediments. Postdepositional mobility introduced by the reduction process has been postulated as a mechanism for a trace element source for elements found in manganese nodules formed on the ocean bottom.

The pH of sea water is affected by temperature, salinity, photosynthesis, and respiration, and by the reaction with the calcium-carbon-oxygen buffering system. Understanding the latter system is commonly acknowledged to be one of the most difficult problems in marine chemistry. Certainly, the problem of CO_2 in the atmosphere and the possible relation of this to its concentration in the ocean is critical to human survival. The reactions can be represented by two sets of equations:

$$CO_2 + H_2O = H_2CO_3 = HCO_3^- + H^+ = 2H^+ + CO_3^{2-}$$

and

$$H_2CO_3 + CaCO_3 = Ca^{2+} + 2HCO_3^-$$

Carbon dioxide in water will form carbonic acid (H_2CO_3), lowering the pH. Carbonic acid will react with calcium carbonate $(CaCO_3)$, resulting in an increase in pH. The H^+ ions formed from the dissociation of carbonic acid are used to form bicarbonate ions $(HCO_3)^-$ from the carbonate ion CO_4^{2-}, derived from the dissociation of $CaCO_3$. Short-term variations are buffered by the presence of the various components. Recall that Chapter 3 discussed how the lifetime of a stream was derived by projecting the time needed to acidify all of the bicarbonate ions $(HCO_3)^-$ (17). Increases in CO_2 through interaction with the atmosphere and through changes in photosynthesis and respiration can be considered as being balanced by the large input of $(HCO_3)^-$ and Ca^{2+} from stream runoff, which would tend to increase pH. In opposition to this is the deposition of carbonates in marine sediments, thereby removing the carbonates from the buffering process and, once again, decreasing pH.

Other chemical elements whose species are capable of accepting a H ion are N, P, and B. Of these, only a boron species, $B(OH)_4^-$, is present in sufficient quantities to be significant—and it supplies less than 5 percent of the total alkalinity. The amount of nitrogen present is high, but is predominantly the relatively inactive N_2 with only minor amounts of NO_3^-, and thus nitrogen is not a factor in pH control. It is probably valid to focus attention on the carbon-oxygen system even though it is evident that an equilibrium condition does not exist everywhere all the time, that kinetics can be critical, and that knowledge of the system may not give us a unique answer.

The reasons for this last statement are that (a) the carbonate system, despite its undoubted short-term buffering of changes in pH, may reflect rather than control the oceanic pH, and that (b) the oceanic pH is actually controlled by equilibrium between dissolved cations and alumino silicates, either suspended or deposited and either carried in from land or formed in the ocean (authigenic). "Weathering" in the sedimentary cycle consumes H or CO_2 with the release of cations, SiO_2, and bicarbonate $(HCO_3)^-$ (18). This relationship has also been presented by Urey (19) as:

$$CaSiO_3 + CO_2 = CaCO_3 + SiO_2$$

where metal silicates react with acid gases to form soluble salts and hydrogen silicates as weathering and solution take place. This argument is not only the key for possible understanding of oceanic pH, but is the heart of the rock cycle. Suppose that CO_2, HCl, SO_2, or any other acid volatile is present in large amounts; then the following sequence could occur—all processes using up H^+: (a) $CaCO_3$ would dissolve, forming $(HCO_3)^-$; (b) clay minerals would exchange ions to pick up H^+ and free Na^+, K^+, etc.; and (c) feldspars and other silicates would break down and form clays and free silica (20). It can also be postulated that the reverse reactions can occur—that is, Na can be added to kaolinite clay to form feldspar. If an equilibrium situation exists, it would be expected that the reactions would take place in both directions.

If the ocean is to remain constant in composition, some means must be found to remove the SiO_2, cations, and bicarbonate formed by the aforementioned reactions. The $(HCO_3)^-$ can be used by organisms, and the SiO_2 could be removed by sedimentation. However, large deposits of SiO_2 are not present on the ocean floor, and the problem of cation removal remains. The mechanism of "reverse weathering" of clay minerals has been proposed. In this process SiO_2, the cations, and kaolinite react to form illite and other "less degraded" clays, with a decrease of pH. The importance of this mechanism, both today and in the geologic past, is one of the chief geochemical disputes today. The major arguments in favor of reverse weathering are that (a) there is no recognizable geologic evidence for the oceans having changed their compositions through time, (b) streams deliver much higher concentrations of $(HCO_3)^-$ and cations than needed to balance the measured cation content in the oceans, (c) cations must be removed in some fashion, and, finally, (d) clay reactions appear to remove the cations. However, if most of the cations never get out of the estuary, the exact mechanism loses some of its importance.

Particulate Matter in the Oceans

Several sources of particulate matter in the oceans have already been discussed in Chapter 3 — river load and airborne dust, with extraterrestrial material (cosmic dust) being only minor. Mechanisms for measuring particulate matter have been of two types: direct sampling, and calculating density from thickness of bottom sediments and from sedimentation rates. The total particulate load in the oceans is estimated to be 10×10^9 tonnes, or an average density of 10-20 ppb, comparable to the weight of Ba dissolved in sea water (Figure 5-2) where Ba is the eighteenth most common element (21). This concentration is 50 to 100 times smaller than previous estimates (22) of 1 ppm. The measurements are biased toward small particles, however, because of their slower settling and longer residence time in the oceans. River load is expected to decrease as surface currents decrease because they can effectively transport only those particles with a settling velocity of less than 0.005 cm/sec, which corresponds to a size of 6-10 μm for spheroids. For the Amazon estuary, as the salinity of the river changes from 0‰ to 10‰, the total suspended load decreases from 500 mg/l (ppm) to 3 mg/l (23).

Other sources of particulate matter are (as mentioned in the previous section) precipitation of compounds from the sea water, particularly $CaCO_3$, and detrital particles, such as shells or fecal matter resulting from biological processes. For the deeper ocean, annual production of biogenic particles exceeds land sources by a factor of 10-20, but the particles of calcite, aragonite, and silica have a tendency to be decomposed. Consequently, surface waters may contain 90 percent biogenic detritus, whereas the bottom sediments contain less than 10 percent biogenic material (24). In surface water, organic matter is 30-70 percent of total particulate matter (TPM) and the hard skeletons

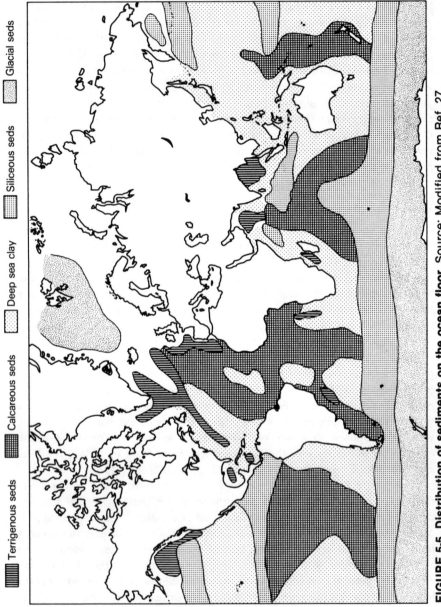

FIGURE 5-5. Distribution of sediments on the ocean floor. Source: Modified from Ref. 27.

composed of $CaCO_3$ and SiO_2 compose 25–50 percent of the TPM (21). There is a pronounced latitudinal effect around the Earth following the pattern of biological productivity.

The degree of decomposition or amount of corrosion of the organic particles is inversely proportional to their initial size. This is because large particles settle faster, and thereby release not only Ca, CO_3, and Si but also any trace elements concentrated by the biological process. For certain sizes (less than 5 μm for calcite and less than 20 μm for radiolarian particles) upwelling can cause a recognizable particle-rich layer above the ocean bottom called a nepheloid layer. The particles reverse their direction but continue to dissolve, creating a narrow bank that is rich in particles in the micrometer-size range (21). Others (25) argue that although the nepheloid layer could be caused by accumulation of particles sinking from water layers at the surface or middle depths, it could also be caused by resuspension of bottom sediments by benthic currents. Increase in Fe, Al, Si, and Mn, as compared with overlying waters, could be caused by preferential loss of particles low in Fe, Al, Si, and Mn (such as $CaCO_3$), but the observed increase in total suspended particles and the variable enrichment of these elements argues against this conclusion.

Ocean-surface suspended matter is also processed by zooplankton, and removed from the surface water by sinking fecal pellets. The particles can be whole or fragmented, dissolution increasing with degree of fragmentation of the pellet. These pellets provide food supply for benthic life, but are assumed to move too rapidly through the water column to absorb soluble ions.

Terrigenous sediment content and sedimentation rate would be expected to be different for each ocean, because of the much greater volume of material drained into the Atlantic and the larger size of the Pacific. On this basis alone, sedimentation rates could be predicted to be 10 times greater in the Atlantic than in the Pacific (26). A first-order generalization could be that an overall erosion rate of 30 mm/1000 yr in the major river basins discussed previously may be compared with a mean deposition rate of 10 mm/1000 yr in the world's oceans, by comparing the relative areas of the two. The sediments formed could also be expected to be influenced by the increased supply of dissolved carbonates from stream runoff, two-thirds of which goes into the Atlantic. The runoff effect is strong, as shown by the distribution of principal types of sediment on the ocean floor, illustrated in Figure 5-5.

One other major factor controls the distribution of carbonate sediments. Below 5000 m depth in the ocean virtually no carbonate is present in particulate form because of the strong degree of undersaturation. The depth at which rates of supply and dissolution of carbonates are equal is termed the carbonate-compensation depth. There is some evidence that the larger particles dissolve while on the bottom, a function of their more rapid transit through the water columns. The lesser mean depth and smaller proportion of

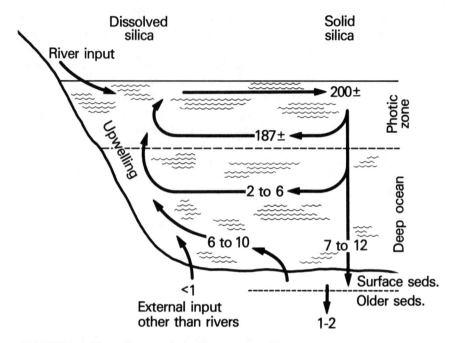

FIGURE 5-6. The silica cycle in the oceans. Figures are in g/m²/yr. Source: Ref. 27. Copyright 1976 by Academic Press, Inc. (London) Ltd. Published with permission.

deep ocean in the Atlantic, compared to the Pacific Ocean basin, should result in a greater amount of carbonate sediments in the Atlantic than the Pacific.

As expected, most of the silica is used and reused by biological organisms close to the surface of the ocean (Figure 5-6). The changes in SiO_2 concentration correspond with changes in phosphate composition. About 4 percent of the skeletal silica formed in the oceans (7–12 g/m² yr) survives to reach the ocean bottom and more than half of this (6–10 g/m² yr) is dissolved. There is a strong correlation between SiO_2 accumulation and organic productivity, not only because SiO_2 is more available, but also because the presence of organic matter appears to stabilize the shells (27). This pattern, coupled with the previously discussed carbonate pattern, can result in the sediment accumulation pattern illustrated in Figure 5-7. Throughout the region shown, sections a–f, clays from all sources are accumulated at a low rate (2 m/1000 yr). As productivity of organisms increases (top curve) there is a region (sections b–e) on both sides of the equator where silica production exceeds dissolution, and silica therefore accumulates in the sediments. The rate is estimated at 4–5 mm/1000 yr. No carbonate accumulates there because the carbonate-

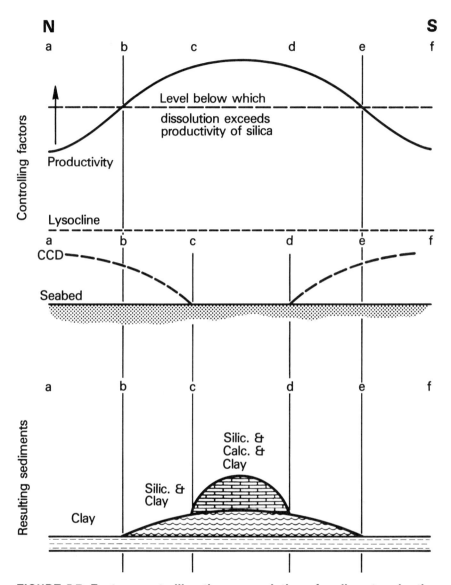

FIGURE 5-7. Factors controlling the accumulation of sediment under the equatorial high-productivity belt. (CCD is carbonate compensation depth). Source: Ref. 27. Copyright 1976 by Academic Press, Inc. (London) Ltd. Published with permission.

compensation depth (CCD) lies above the ocean floor, and all the carbonate dissolves. In regions c and d, however, the CCD is depressed under the region of high productivity, so carbonate is deposited at a rate of 10–20 mm/1000 yr. The bulge of thick sediments should be displaced vertically as the Pacific plate moves northward. There is some evidence for this (26).

The movement of ocean water itself from ocean to ocean can also be a major control (28). As Davies and Gorsline stated (27, p. 67):

> If the model is modified to the three box version shown in [Figure 5-8] several important consequences result. Firstly, because the Atlantic is the only ocean extending into the Arctic it is also the only ocean which has strong western boundary currents in the northern hemisphere. Secondly, because of the displacement of the land masses into the northern hemisphere it is possible to develop circumpolar flow around Antarctica, and in fact the Circum Antarctic Current transports the largest volume of water of any current in the ocean. Cold, deep ocean water originates in the Norwegian Sea, the Weddell Sea and the Ross Sea, and the Circum-Antarctic waters act as a mixing and recooling region. About half the deep water originates in the Norwegian Sea (Broecker, 1974) [29], and there is a net flow of deep water from the Atlantic into the Pacific and Indian Oceans which results in a strong horizontal segregation of nutrients. The accumulation of silica in the Pacific Ocean results from the vertical concentration gradients of silica in ocean water, and is a direct result of a steady pumping of silica-rich deep water from the Atlantic to the Pacific and of surface water depleted in silica from the Pacific to the Atlantic (Berger, 1970) [28]. The reverse effect is observed with carbonate because carbonate dissolution increases with pressure, with decreasing temperature, and with increasing total carbon dioxide concentration. Cold ocean water is rich in oxygen when at the surface, but becomes gradually depleted of it as a result of biological processes. As a consequence of the decrease in oxygen content there is a gradual increase in the total carbon dioxide concentration. Thus, the deep waters reaching the Pacific are much more corrosive to carbonate than they were soon after sinking in the Atlantic.

The amount of suspended sediment transported through interoceanic exchange has been estimated to approximate the total load (dissolved and particulate) delivered to the oceans by rivers (22). Of this amount, 98 percent moves eastward around Antarctica and appears to be accumulating slowly in the Pacific.

There are increasing work in trace-element analysis of sediments and efforts to determine the detrital fraction from the authigenic fraction. Elements independent of sedimentation rate would be expected to have very little authigenic concentration, and to be concentrated in the detrital fraction; Sc, Ti, and Th follow this pattern (30). A model of uniform authigenic deposition, superimposed on a varying background of detrital input, would produce a

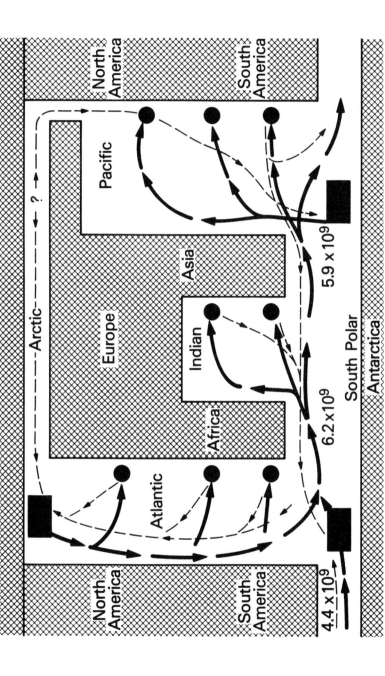

FIGURE 5-8. Ocean circulation. Heavy lines are bottom currents. Light lines are surface currents. Upwelling is indicated by circles and downwelling by squares. Numbers indicate relative movement (km³/yr). Sources: Modified from Refs. 27, 29.

negative correlation with sedimentation rate; Mn, Co, Ni, Fe, and Cu follow this pattern, with 90 percent Mn, 80 percent Co and Ni, and 50 percent Cu being authigenic. The roles of absorption, underwater volcanic addition, loss or addition in estuaries, etc., complicate this analysis.

Ocean Sediments

Sediments that reach the ocean bottom through all of the processes with all of the reactions previously discussed are obviously not simple and straightforward, nor are they easy to analyze, classify, or understand. However, the processes that today operate to change these sediments through chemical reaction and compaction into sedimentary rocks are some of the most interesting. The movement of organic remains to form fossil-fuel deposits, the movement of Fe and Mn to form nodules, and the enrichment of phosphorus or copper or sulfides are of economic interest as well.

Material transported to the ocean by streams, by ice, or through the atmosphere is termed lithogenous or rock-originated. Some of this material does not react with sea water, but passes through the water column and is deposited after possibly adsorbing organics and cations. Hydrogenous (authigenic, or ocean-originated) material in marine sediments is formed in sea water by inorganic reactions, either by precipitation of minerals from sea water (primary) or by modification of lithogenous material (secondary). Hydrogenous materials are classified either as halmyrolysates, which are formed as a result of reactions between sediment components and sea water, or (b) precipitates, which are primary inorganic compounds formed by direct removal of elements from sea water, with pre-existing sediments having no active role (31). Halmyrolysis is thus the process operative between weathering (breakdown of the rock into sediment) and diagenesis (consolidation of the sediments into rock). The process of precipitation occurs when sea water is saturated with respect to a particular mineral—usually only $CaCO_3$, but saturation can be reached for other minerals either through evaporation of sea water or through changes in oxidation potential so that the normally oxidized species are reduced. The most important hydrogenous materials in marine sediments are shown in Figure 5-9.

Particles from land sources can absorb material, as can authigenic minerals, thus clouding any analysis by mineral type. In addition, interstitial waters in ocean-floor sediment can have significantly different compositions than overlying sea water (32). Proper evaluation of this process currently has been delayed by the recent realization (33) that rinsing of clays with distilled water during the analytical procedure can give results that are the opposite of what existed naturally—that is, Na^+ rather than Mg is the major exchangeable cation for bound Ca^{++}.

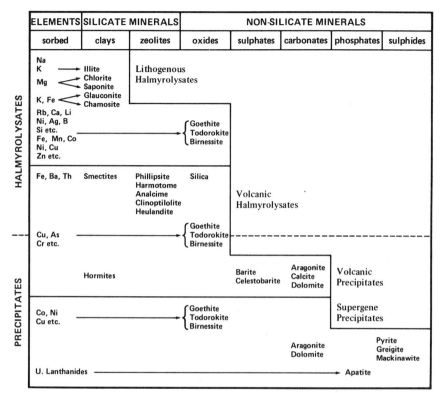

FIGURE 5-9. A classification of the hydrogenous material in marine sediments. Source: Ref. 31. Copyright 1976 by Academic Press, Inc. (London) Ltd. Published with permission.

The major types of oceanic particulate matter combine as sediments in different proportions in different parts of the ocean floor. What results is two major types of deep-sea sediments — "clays" and "carbonates." The clays, so called because of the small size of their constituent particles rather than their mineralogy, are composed of lithogenous particles and halmyrolysate particles that have been deposited at a slow rate. Some precipitates, especially ferromanganese particles, are also present. The carbonates have a faster rate of deposition, and contain significant amounts of shell fragments. The bulk analyses of these major types are given in Table 5-3 (34).

At the surface of most ocean sediments is a mixed layer of fine-grained hydrated particles and organic matter. This ooze contains living organisms that use the organic matter as a source of nutrients. By degrading the ooze, these organisms destroy its colloidal aspect and aid in its consolidation as sediment. The reactions proposed (16) were mentioned previously. Recent

Element	Deep-sea carbonate	Deep-sea clay
Li	5	57
Be	X	2.6
B	55	230
Sc	2	19
V	20	120
Cr	11	90
Mn	1000	6700
Fe	9000	65000
Co	7	74
Ni	30	225
Cu	30	250
Zn	35	165
Ga	13	20
Ge	0.2	2
As	1	13
Se	0.17	0.17
Rb	10	110
Sr	2000	18
Y	42	9
Zr	20	150
Nb	4.6	14
Mo	3	27
Ag	X	0.11
Cd	0.23	0.21
In	0.02	0.08
Sn	X	1.5
Sb	0.15	1.0
Cs	0.4	6
Ba	190	2300
La	10	115
Ce	35	345
Pr	3.3	33
Nd	14	140
Sm	3.8	38
Eu	0.6	6
Gd	3.8	38
Tb	0.6	6
Dy	2.7	27
Ho	0.8	7.5
Er	1.5	15
Tm	0.1	1.2
Yb	1.5	15
Lu	0.5	4.5
Hf	0.41	4.1
Re	0.004	0.001
Hg	0.46	0.32
Tl	0.16	0.8
Pb	9	80
Th	—	5
U	—	1

TABLE 5-3. Concentrations of some trace elements in deep-sea sediments (values in ppm). Order of magnitude estimates are indicated by X.

Source: Ref. 34. Copyright 1976 by Academic Press, Inc. (London) Ltd. Published with permission.

measurement of pore-water nitrate concentrations (35) has shown an increase from bottom-water values downward to a depth of 5 cm, and then a decrease to undetectable levels apparently as shallow as 40 cm. These concentrations and gradients were ascribed to zones of oxidation of organic material by free oxygen, nitrate ion, and sulfate ion, with sulfate oxidation taking place only after the nitrate is exhausted. Calculated rates indicate that sediments are an important site for marine denitrification. The depth of this major change from an oxidizing to a reducing environment can be 1 to 5.5 mm in, for example, a coastal marine sediment (36). The elements certainly change their behavior through these intervals with the decrease in oxidation potential. Indeed, the migration of Cd, Pb, and Zn is from sediment to pore water under oxidizing conditions, regardless of type of sediment (37). This supports the earlier discussion that emphasized elements would tend to stay in solution, and in their most oxidized form, in normal sea-water conditions. It also emphasizes the conclusion that unless there is a complete absence of organic matter one would expect to find a sharp increase in reducing conditions and an increased absorption of heavy elements.

Much recent interest has focused on estimating the chemical contribution from the various sediment sources (38) to the ocean bottom sediments. One consequence is that authigenic sediment sources are much less abundant than previously estimated. A second consequence is the awareness that more than one source is often involved; mixtures of biogenous rich shale, East Pacific Rise sediments, and basaltic material show excellent agreement with the composition of deep-sea sediments found (39, 40). A third consequence is that the importance of biological material as a constituent of pelagic sediment has now been recognized.

A similarity exists between the distribution pattern of opaline silica accumulated and that of Fe, Mn, Ca, Ni, and Co, with a remarkable regularity in the minor- and trace-element ratios among the lower marine animals when the analyses are normalized (Al + Fe + Zn = 1) (39). Organisms apparently act without discrimination during feeding for these elements, even though the alkaline and alkaline earth elements are rejected by the physiological reactions of the organisms. Taxonomic effects appear to be negligible, and those organisms that concentrate a particular element are present either in very minor amounts or in constant proportions. Plankton in the central Pacific have much higher concentrations of minor and trace elements than those closer to the continents—which could mean (a) that once the elements have been incorporated into the food chain, they tend to remain there and become increasingly concentrated, or (b) with more nutrients present close to the continents, less water must be processed for feeding and fewer elements extracted, or (c) a combination of both. The distribution of Cu and Zn appear to be most strongly related to biogenous material but Ni, Al, Fe, and Mn are also concentrated this way (38).

Until recently the role or mechanism of hydrothermal sedimentation from volcanic sources, although not known exactly, was used to explain why deep-ocean sediments are richer in certain elements (Mn, Co, Ni, and Zn) than near-shore sediments. Other explanations included a "differential transport" of the colloid/clay-sized river particles—rich in metals adsorbed while the particles were in fresh water—to the deep ocean, with sedimentation there (41). This differential transport seems unlikely, in view of our discussion of estuarine processes and in recognition that the different oceans receive completely different volumes of sediments. Sediment thickness in the Atlantic is greater than 1 km, in the Pacific it is much less than that, and the world average is about 0.5 km (34).

An alternative "trace element veil" theory was proposed (42), which involved the precipitation of trace elements by superimposing them at a constant rate on suspended particles being deposited at various rates. Thus, the deep-sea clays of the Pacific would be richer in trace elements than those in the Atlantic because the Pacific sediments are deposited at a slower average rate. Atmospheric transport of land particles, volcanic additions, and a nonhomogeneous distribution can cloud this relationship. (For a more thorough discussion of these arguments, see Ref. 34, p. 365 f.)

The magnitude and importance of hydrothermal sources has been partially clarified with the recent discovery of hydrothermal vents releasing water at almost 400°C along the East Pacific Rise (43). These vents are actively forming Fe-rich sulfide mineral deposits. Similar nearby deposits have been sampled and shown to be essentially Fe and Zn minerals with less abundant Cu-rich minerals (44). The source of sulfur could be either from sulfates in sea water or from a sulfur-rich melt beneath the ridge system. In the first case, the sulfate is reduced when the sea water, percolating downward, reacts with the Fe of the pre-existing basalt. The sea water changes from an alkaline to a slightly acid solution and, at temperatures of about 400°C, can leach Cu and Zn. The elements would be deposited when the now-hydrothermal solution is cooled on the ocean bottom. In the second case, heated sea water reacts with the magma system (whether fluids, minerals, or both is unknown) and can produce the metal-rich solutions that are observed.

Ferromanganic nodules are most closely related to underwater processes at "hot spots" on the ocean bottom, possibly through leaching of underwater lavas coupled with the changes in oxidation potential previously discussed. It seems apparent that the presence of Fe-Mn nodules is closely tied to the relationship between the "oxygen zero" boundary and the sediment-sea water interface (31).

The literature of ferromanganese nodules is extensive (e.g., 45, 46). The basic mineralogy was previously indicated: goethite, an iron oxyhydroxide,

TABLE 5-4
Elements Adsorbed on Marine Fe-Mn Nodules

Greatly Enriched	Moderately Enriched With Fe	Moderately Enriched With Mn	Slightly Enriched	Probably Not Enriched
Thalium	Boron	Zinc	Iron	Cadmium
Molybdenum	Nickel	Barium	Ytterbium	Titanium
Manganese	Copper		Mercury	Gallium
Cobalt	Silver			Vanadium
	Cadmium			Strontium
	Lead			Yttrium
				Zirconium

Source: See Table 3-7, also Refs. 45, 47 and 48.

FeOOH; todorokite, a $Mn^{+2}-Mn^{+4}$ mixed oxide with roughly a Mn_3O_5 ratio; and birnessite, a defect Mn^{4+} structure, δMnO_2. The different oxidation-state radii of the ions allow different geochemical substitutions, with Ni, Cu, and Zn enriched in todorokite, and Co and Pb enriched in birnessite. Because Pb does not appear to be adsorbed on Mn oxides, it must substitute as Pb^{4+} (Table 5-4). With the exception of the strong enrichment of B and Mo in the marine nodules, the enrichment factors for marine nodules relative to crustal enrichments generally agree with what was derived from fresh-water oxide coatings (see Table 3-7). These two elements occur as anionic complexes in sea water in a relatively enriched state, and can be included in the oxides as anionic adsorbants. Nearshore nodules have a lower oxidation state of manganese, which can approach a manganese to oxygen ratio of 2 to 3. The trace-element content is much lower in the Atlantic relative to the Pacific, indicating perhaps a much greater growth rate.

The sediments of the ocean floor, far from being a sink for material transported into the oceans by streams or through the atmosphere, are part of a dynamic process, the scale of which we are just beginning to comprehend. We are not yet able to determine the amount of sea water that is cycled through the hydrothermal system along mid-ocean ridges. We are not yet able to delineate the contribution of the different sediment sources to a particular sediment although we are getting close (49). We do not know the mechanisms of concentration of metals by biogenous material, but we can show a relationship. We know that the ocean floor is comparable to the soil reservoir in the number and variety of geochemical reactions that occur. We expect this parallel to be more obvious as we increase our sampling of the ocean bottom—as we learn what questions to ask of the oceans.

References Cited

1. Brewer, P. G., 1975, Minor Elements in Sea Water; p. 416-495 *in* Chemical Oceanography, v. 1, G. A. Riley and G. Skirrow, eds., Academic Press, London, 606 p.
2. Burton, J. D., 1977, The Composition of Sea Water; Chemistry and Industry, no. 14, p. 550-557.
3. Pytkowicz, R. M., and D. R. Kester, 1971, The Physical Chemistry of Sea Water; Oceanography and Marine Biology Annual Review, v. 9, p. 11-60.
4. Riley, J. P., 1975, Analytical Chemistry of Sea Water; p. 193-542 *in* Chemical Oceanography, v. 3, 2nd ed., J. P. Riley and G. Skirrow, eds., Academic Press, London, 564 p.
5. Turekian, K. T., 1969, The Oceans, Streams and Atmosphere; p. 297-323 *in* Handbook of Geochemistry, v. 1, K. H. Wedepohl, ed., Springer-Verlag, New York, 442 p.
6. Wilson, T.R.S., 1975, Salinity and the Major Elements of Sea Water; p. 365-411 *in* Chemical Oceanography, v. 1, 2nd ed., J. P. Riley and G. Skirrow, eds., Academic Press, London, 606 p.
7. Turekian, K. T., 1977, The Fate of Metals in the Oceans; Geochimica et Cosmochimica Acta, v. 41, p. 1139-1144.
8. Goldberg, E. D., ed., 1975, Chemical Speciation in Seawater; p. 17-41 *in* The Nature of Seawater, Dahlem Workshop, Dahlem Konferenzen, Berlin, 719 p.
9. Holland, H. D., 1978, The Chemistry of the Atmospheres and Oceans; Wiley, New York, 351 p.
10. Stumm, W., and P. A. Brauner, 1975, Chemical Speciation; p. 174-236 *in* Chemical Oceanography, v. 1, 2nd ed., J. P. Riley and G. Skirrow, eds., Academic Press, London, 606 p.
11. Ehrehardt, M., 1977, Organic Substances in Sea Water; Marine Chemistry, v. 5, p. 307-316.
12. Sharp, J. H., 1973, Total Organic Carbon in Seawater—Comparison of Measurements Using Persulfate Oxidation and High Temperature Combustion; Marine Chemistry, v. 1, p. 211-229.
13. Chester, R., and J. H. Stoner, 1974, The Distribution of Particulate Organic Carbon and Nitrogen in Some Surface Waters of the World Ocean; Marine Chemistry, v. 2, p. 263-275.
14. Pocklington, R., 1977, Chemical Processes and Interactions Involving Marine Organic Matter; Marine Chemistry, v. 5, p. 479-496.
15. Garrels, R. M., F. T. Mackenzie, and C. Hunt, 1975, Chemical Cycles and the Global Environment; Wm. Kauffman, Los Altos, Calif., 205 p.
16. Deuser, W. G., 1975, Reducing Environments; p. 1-37 *in* Chemical Oceanography, v. 3, 2nd ed., J. P. Riley and G. Skirrow, eds., Academic Press, London, 564 p.
17. Odén, S., 1976, The Acidity Problem—An Outline of Concepts; Water, Air, and Soil Pollution, v. 6, p. 137-166.
18. Sillin, L. G., 1967, Gibbs Phase Rule and Marine Sediments; p. 57-69 *in*

Equilibrium Concepts in Natural Water Systems, Advances in Chemistry no. 67, American Chemical Society, Washington, D.C., 344 p.

19. Urey, H. C., 1956, Regarding the Early History of the Earth's Atmosphere; Geological Society of America Bulletin, v. 67, p. 1125-1128.

20. Krauskopf, K. B., 1967, Introduction to Geochemistry; McGraw-Hill, New York, 721 p.

21. Lal, D., 1977, The Oceanic Microcosm of Particles; Science, v. 198, p. 997-1009.

22. Lisitzin, A. P., 1972, Sedimentation in the World Oceans; Society of Economic Paleontologists and Mineralogists Special Publication no. 17, Tulsa, Okla., 218 p.

23. Sholkovitz, E. R., and Price, N. B., 1980, The Major Element Chemistry of Suspended Matter in the Amazon Estuary; Geochimica et Cosmochimica Acta, v. 44, p. 163-171.

24. Drake, D. E., 1976, Suspended Transport and Mud Deposition on Continental Shelves; p. 127-158 *in* Marine Sediment Transport and Environmental Management, D. Stanley and D. Swift, eds., Wiley, New York, 602 p.

25. Baker, E. T., and R. A. Feely, 1978, Chemistry of Oceanic Particulate Matter and Sediment: Implications for Bottom Sediment Resuspension; Science, v. 200, p. 533-535.

26. Stoddart, D. R., 1969, World Erosion and Sedimentation; p. 43-64 *in* Water, Earth, and Man, R. J. Chorley, ed., Methuen, London, 588 p.

27. Davies, T. A., and D. S. Gorsline, 1976, Oceanic Sediments and Sedimentary Processes; p. 1-79 *in* Chemical Oceanography, v. 5, 2nd ed., J. P. Riley and R. Chester, eds., Academic Press, London, 401 p.

28. Berger, W. H., 1970, Biogenous Deep-Sea Sediments: Fractionation by Deep-Sea Circulation; Geological Society of America Bulletin, v. 81, p. 1385-1402.

29. Broecker, W. S., 1974, Chemical Oceanography; Harcourt-Brace-Jovanovich, New York.

30. Krishnaswami, S., 1976, Authigenic Transition Elements in Pacific Pelagic Clays; Geochimica et Cosmochimica Acta, v. 40, p. 425-434.

31. Elderfield, H., 1976, Hydrogenous Material in Marine Sediments, Excluding Manganese Nodules; p. 137-215 *in* Chemical Oceanography, v. 5, 2nd ed., J. P. Riley and R. Chester, eds., Academic Press, London, 401 p.

32. Price, N., 1976, Chemical Diagenesis in Sediments; p. 1-59 *in* Chemical Oceanography, v. 6, 2nd ed., J. P. Riley and R. Chester, eds., Academic Press, London, 414 p.

33. Sayles, F. L., and P. C. Mangelsdorf, Jr., 1977, The Equilibration of Clay Minerals With Seawater: Exchange Reactions; Geochimica et Cosmochimica Acta, v. 41, p. 951-960.

34. Chester, R., and S. R. Aston, 1976, The Geochemistry of Deep-Sea Sediments; p. 280-390 *in* Chemical Oceanography, v. 6, 2nd ed., J. P. Riley and R. Chester, eds., Academic Press, London, 414 p.

35. Bender, M. L., K. A. Fanning, P. N. Froelich, G. R. Health, and V. Maynard, 1977, Interstitial Nitrate Profiles and Oxidation of Sedimentary Organic Matter in the Eastern Equatorial Atlantic; Science, v. 198, p. 605-609.

36. Revsbech, N. P., B. B. Jorgensen, T. H. Blackburn, 1980, Oxygen in the Sea Bottom Measured with a Microelectrode; Science, v. 207, p. 1355-1356.

37. Cobler, R., and J. Dymond, 1980, Sediment Trap Experiment on the Galapagos Spreading Center, Equatorial Pacific; Science, v. 209, p. 801-803.

38. Leinen, M., and D. Stakes, 1979, Metal Accumulation Rates in the Central Equatorial Pacific During Cenozoic Time; Geological Society of America Bulletin, v. 90, p. 357-375.

39. Boström, K., O. Joensuu, and I. Brohm, 1974, Plankton: Its Chemical Composition and its Significance as a Source of Pelagic Sediments; Chemical Geology, v. 14, p. 255-271.

40. Boström, K., L. Lysen, and C. Moore, 1978, Biological Matter as a Source of Anthigenic Matter in Pelagic Sediments; Chemical Geology, v. 23, p. 11-20.

41. Turekian, K. T., 1967, Calculations; p. 227-244 in Progress in Oceanography, v. 4, M. Sears, ed., Pergamon Press, Oxford, 344 p.

42. Wedepohl, K. H., 1960, Spurenanalytische Untersuchugen an Tiefsectonen aus dem Atlantik; Geochimica et Cosmochimica Acta, v. 18, 200-231.

43. Rise Project Group: 1980, East Pacific Rise: Hot Springs and Geophysical Experiments; Science, v. 207, p. 1421-1433.

44. Hekinian, R., M. Fevrier, J. Bischoff, P. Picot, W. Shanks, 1980, Sulfide Deposits from the East Pacific Rise Near 21°N; Science, v. 207, p. 1433-1444.

45. Cronan, D. S., 1976, Manganese Nodules and other Ferro-Manganese Oxide Deposits; p. 217-263 in Chemical Oceanography, v. 5, 2nd ed., J. P. Riley and R. Chester, eds., Academic Press, London, 401 p.

46. Mielke, J. E., 1976, Ocean Manganese Nodules, 2nd ed.; Committee Print, Senate Committee on Interior and Insular Affairs, U.S. Congress, 94th, 1st sess., February, 163 p.

47. Glasby, G. P., 1974, Mechanisms of Incorporation of Manganese and Associated Trace Elements in Marine Manganese Nodules; Oceanography and Marine Biology Annual Review, v. 12, p. 11-40.

48. Parks, G.A., 1975, Adsorption in the Marine Environment; p. 241-308 in Chemical Oceanography, v. 1, 2nd ed., J. P. Riley and G. Skirrow, eds., Academic Press, London, 606 p.

49. Leiner, M., and N. Pisias, 1980, Geochemical Partitioning: Application of an Objective Technique for End-Member Characterization; Geological Society of America Abstracts With Programs, 1980, p. 470-471.

6
Biota and the Biosphere

Life does not proceed by the association and additions
of elements but by dissociation and diversion.
—H. Bergson

Introduction

The biosphere (Figure 6-1) has been defined in two major ways: first, as "that portion of the earth's crust which is populated by organisms and is characterized by geologic activity of all forms of living matter—plants, animals, and microorganisms" (1, p. 15); and second, as "the total mass of living organisms" (2, p. 42; 3). The second definition would not include the organic matter produced in the soil processes discussed in the previous chapter, nor would it include any other remnants of organic activity such as coal and oil. In order to discuss geochemical processes operative through living organisms, we will distinguish among *biosphere* as the place where living matter interacts with the other earth systems, *biomass* as the mass of living organisms, and *biota* as a general descriptive term for living organisms.

The distribution of the biosphere is very uneven both horizontally and vertically at the surface of the Earth, with land plants dominating the biomass. Only small concentrations of birds, pollen, and insects extend above the tops of trees in lowland areas, and temperature and elevation have set drastic limits on biota in mountains as shown by the well-known "tree line." The upper limit of the biosphere is considered to be controlled by the permanent freezing temperature of water. The downward extension of biota is also drastically limited. Microorganisms are active only in the uppermost soil zones. Roots can extend downward tens of meters, but they make up only a small volume of the total biomass. Microorganisms have been found in deep-seated water associated with petroleum deposits and are presumably limited by the boiling temperature of water.

With these limits, one would expect to find that the oceans are completely within the biosphere. Indeed, we do find life at all depths within the oceans,

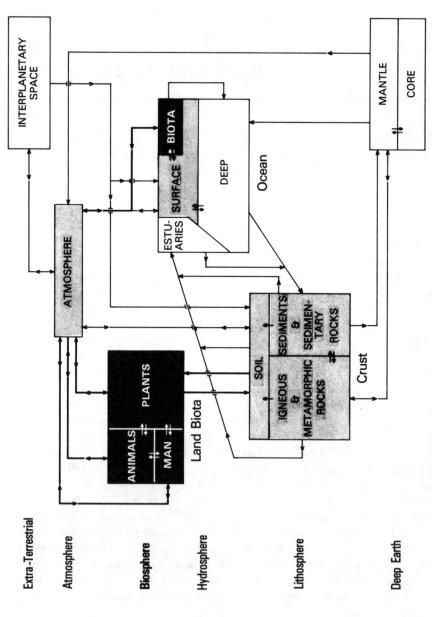

FIGURE 6-1. **The biota reservoir in the geochemical cycle.** Reservoirs are in black, fluxes are given by heavy lines. Shaded reservoirs are those with major interactions with the biota reservoir.

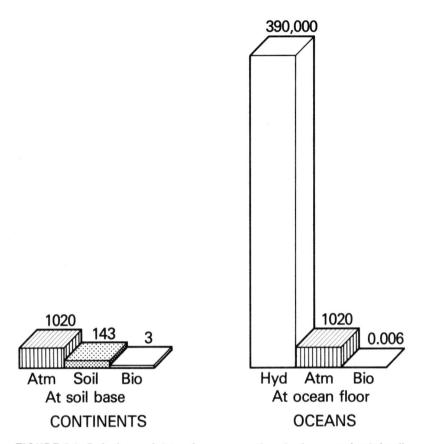

FIGURE 6-2. Relative weights of some geochemical reservoirs (g/cm^2).

although, as with the land, its distribution is highly uneven. For example, dense animal populations have recently been found associated with hydrothermal vents on the ocean floor (4, 5), but other areas of the ocean bottom are virtually lacking in living things.

We can consider the soil-atmosphere interface and the ocean-atmosphere interface to be regions where processes of formation of biota predominate. On the other hand, due to the processes discussed in the next section, the bottom of the soil and the lower limit of the ocean are regions where these organic compounds are broken down and their products accumulate.

The relative masses of the major reservoirs are indicated in Figure 6-2. The weight of biota in the biosphere is seen to be small in relation to the other geochemical reservoirs. The impact of the biosphere on the geochemical movement of elements is volumetrically significant and, of course, is of vital human interest.

We hear much these days of the running down of the Universe, of increases in entropy and in randomness, of transformations of useful energy into wasted heat, of the general hopelessness that awaits. The discussion that follows emphasizes that order is created out of randomness and that the very existence of biota indicates the miracles of life from the geochemical view. Life exists, and in existing, perpetuates itself. How this process is part of the overall world geochemical processes is the subject of the rest of this chapter.

Composition

In the introductory chapter of this book we discussed the principal components of living matter, the macronutrients, and the essential trace elements. The first part of this chapter will deal with those major components. Unfortunately, information is sadly lacking and exceptionally difficult to measure. We must know the composition and amounts of each chemical species in order to get a true value—but the rate of change is such that that value probably would not be valid by the time we finished measuring it. Let us begin by describing, in a roughly semiquantitative manner, the elements of interest.

As stated in earlier chapters of this book, the major constituent of biota is water, ranging from about 50 percent in woody trees to 99 percent in jellyfish (6). The water in the interior of cells can amount to 50 percent of an animal's body weight, as shown in Figure 6-3. Extracellular water can account for another 15 percent. Cellular water can neither boil nor freeze without the death of the organism. This sets the limits of the biosphere in the physical world. A single class of compounds, hydrate of carbon (or carbohydrate) and its polymers $(CH_2O)_n$, which include cellulose, can make up more than 98 percent of the dry body (7). In some carbohydrates and in proteins, sulfur and nitrogen are present. Phosphorus, although not present in the compound, is essential in regulating its formation. Several different estimates of the ratios of these six elements are given in Table 6-1.

Several differences between marine and land biota can be seen. The carbon/nitrogen (C/N) ratio is lower for marine biota than for land biota. This probably reflects the increased volumetric importance of proteins relative to carbohydrates in marine biota. One would also expect the proteins to give a lower carbon/sulfur (C/S) ratio and, generally, marine biota do have a significantly lower C/S ratio than do land biota. As the analyses of marine proteins indicates, however, there are clearly other factors involved. The lower C/N ratios for land biota in the tabulation are based on analyses done before about 1960 (2, 3).

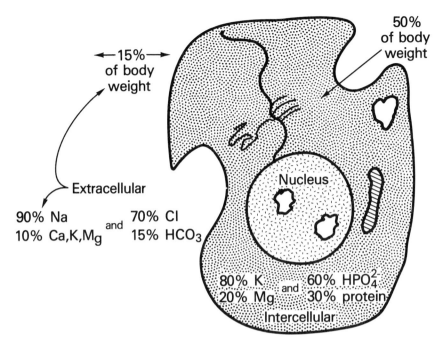

FIGURE 6-3. Composition of extracellular and intercellular fluids. Source: Information from Ref. 6.

Soil humus and organic matter in sedimentary rocks have been suggested as having elemental ratios similar to that of marine biota (9, p. 67). Some tabulations, however, show soil humic acid C/N ratios of 11 to 18 (10), which is closer to land biota values. One would expect very little agreement between the C/N ratio of marine biota and that of organic matter in sedimentary rocks. This organic material changes composition with time, temperature, and depth of burial, in the process losing CO_2, H_2O, and CH_4. In extreme cases the end product is graphite or coal (11). In the production of coal from biota, there is a general increase in nitrogen content. Neither lignin nor cellulose contain nitrogen, so it has been suggested that this change in C/N ratio reflects the activity of (decomposing) bacteria (6). In any event, the elemental ratios do not appear to have much predictive use other than to show that the major element composition of land biota is $(CH_2O)_n$ and that of marine biota is $(CH_2O)_{106}(NH_3)_{16}H_3PO_4$.

The carbohydrates serve both a structural function (such as cellulose and chitin serving as mechanical supports) and a nutrient function (such as starch providing an energy source for metabolism). Three broad types of metabolism are carried out in the biosphere. The primary metabolic function is

TABLE 6-1
Major Element Atomic Ratios in Biota

	Hydrogen	Oxygen	Carbon	Nitrogen	Phosphorus	Sulfur	C/N	C/S	Reference
	310	160	230	13	–	1	18	230	2
Land	2960	1480	1480	16	1.8	1	90	1500	7
Biota	2970	1480	1480	16	0.8	1	90	1500	8
	1600	800	800	9	1	5	90	160	9, p.67
	560	280	280	19	1	1	15		9, p.106
Marine	263	138	106	16	1	1	7		6
Biota	212	106	106	16	1	2	7	50	9
	*300	62	190	56	1	1	3.3	190	6, p.395
	**1066	120	618	5	7	1	134	618	6, p.395

* Marine protein
** Marine lipids

photosynthesis, where solar radiation provides the energy needed to reduce carbon dioxide, form these carbohydrates, and free oxygen. The reaction is

$$nCO_2 + 2nH_2A + E = (CH_2O)_n + nA_2 + nH_2O$$

where H_2A (the hydrogen donor) may be H_2S for some bacteria, H_2O as in blue-green algae and higher plants, or organic compounds as in purple bacteria; and E indicates the addition of energy. Because more than 99 percent of the biomass is in higher plants, photosynthesis must be considered the key process—although from the standpoint of humans, one might argue that oxygen respiration is also reasonably important. The dense animal populations surrounding recently discovered hydrothermal vents on the East Pacific sea-floor are apparently based on a chemosynthesis rather than photosynthesis (5). There the oxidation of the reduced sulfides emitted from the vent apparently releases enough energy to reduce the carbon dioxide to organic matter.

In respiration, the second major type of metabolism, molecular oxygen is used to oxidize organic compounds to CO_2 and H_2O, by which the organism derives energy and synthesizes the necessary structural building blocks. The third type of metabolism is fermentation, whereby the organisms reduce oxidized forms of organic matter, liberating CO_2 and a variety of organic compounds such as alcohols and amines. Photosynthetic organisms are restricted to shallow environments (sunlight is necessary), respiratory organisms to oxygenated environments (free oxygen is necessary), and fermenters to oxygen-free environments. All three processes operate throughout the considerable range of naturally occurring conditions of acidity and oxidation potential previously discussed.

The geochemical behavior of the major elements generally follows that of water. Carbon, nitrogen, and sulfur form compounds that are water-soluble and move with the liquid water. These elements also form volatile compounds such as CO_2, CH_4, NH_3, H_2S, SO_2, and N_2O, which can move through the atmosphere as water vapor does. These elements follow the hydrologic cycle and are controlled by it, as discussed earlier. Indeed, the combination of solubility and volatility of the five major elements explains their mutual interaction to form biota.

If an element is not volatile, it is washed from the biosphere and lithosphere into the ocean by rainfall and runoff. The element does not normally return to the cycle through the atmosphere, so the oceans become a sink—certainly within 100,000 years. The flux is not uniform but has many interruptions, with the biosphere providing the major reservoir—or at least the only chance for a closed natural cycle. The prospect of the ocean as a sink also indicates why many elements (e.g., vanadium, cobalt, nickel, and molybdenum) are best known in marine biota and cycle mainly within the ocean.

There is another major implication in the behavior of nonvolatile elements. If the crust-soil reservoir contains a liberal amount of an element, or if the flux to the biota reservoir is large, the biota can take what is needed and waste the rest. On the other hand, if the quantity or the flux is small, the element will be in short supply. If such a supply is stable, the output of the entire system can adjust to the rate of exploitation of this element. As an analogy, the performance of a bureaucracy can be closely geared to the availability of paper clips (7) or a copy machine (8).

Phosphorus, for example, is a water-soluble but nonvolatile element that accumulates in oceanic sediments. Phosphates are mined and put back into the cycle, but phosphorus does not continue to move through the geochemical cycle naturally. For this reason, and because the rate of accumulation in concentrations great enough for mining is much less than the rate of consumption of phosphate for fertilizer, the future availability of phosphate is seen by some to be the limiting factor for agricultural production.

The second major compositional group of elements in biota includes elements that are soluble in water and present in the cellular and extracellular water, as mentioned previously. Potassium, sodium, magnesium, and calcium are the major cations; the major anions are chlorine, bicarbonate, sulfate, and the phosphates. The distribution of these elements in intercellular fluids and extracellular fluids is very uneven, as Figure 6-3 indicates. All forms of life depend on these cations. The cell walls (reinforced in plants by a layer of cellulose) act as membranes, and control the movement of the ions. Biological species that have poor control mechanisms have restricted access to different environments with the result that their evolution is limited (12). Those species with mobile environments have no such environmental restriction but instead are dependent on the soundness of their cell-wall membranes. Calcium is particularly important in controlling this permeability (13). Therefore, the recent findings that indicate a systematic loss of calcium with weightlessness in humans in outer space has consequences regarding just how far we can extend our own environment (14).

Exact reasons for the concentration of K^+ inside and Na^+ outside the cell are unknown. One possible explanation is tied to the expulsion of Ca^{2+} (the same ionic radius as Na^+) from inside the cell, thus prohibiting crystallization of calcium carbonate, calcium phosphate, and calcium sulfate. The potentially crystallizing material can be transported by the extracellular fluids to a concentration that will permit the formation and growth of eggshells, skeletons, or other structural units. Potassium appears to be especially important as a stabilizing ion in the organization of protein (13) and therefore must be available within the cell.

Failure of the cell-wall membrane to transmit Ca ions could lead to crystal growth and possible deposition, as in cataracts, stones, cartilage, and hardening of soft tissues and arteries in humans (12). Extracellular effects of Ca involve digestion enzymes and blood clotting, among others. Within the cell, muscle contraction results from the activation of enzymes that occur when Ca is introduced. Nerve cells act mainly as conduits for cations, where the reaction occurs by the abnormal acceptance of Na^+ instead of rejection. Magnesium within the cell appears to occupy a balancing position, moving in and out so as to balance the effects of Ca and Na. Effects and processes involving the potassium, sodium, magnesium, and calcium cations both within and outside the cell are still not defined, but there is no question that they occupy a critical area of biochemistry.

The relative proportions of the major cations do not remain constant with plant type, as shown in Figure 6-4. The general decrease in concentration of these cations with increased complexity of plant type is balanced by the increasing significance of the minor elements. There is no apparent correlation of concentration of the major cations in plants with their concentrations in the crust, the soil, or in sea water. If life began in the oceans, as some have postulated, we would expect that extracellular fluids would give a reasonable estimate of the composition of those oceans. The lack of obvious correlation with whole-plant analyses could thus be expected. An extension of this argument has given rise to the popular belief that mammalian blood plasma is really dilute sea water—but it is not. The ratio of element concentration in sea water to its concentration in plasma (both values normalized to the concentration of Mn) ranges downward from 30 for Ca to 24 for Na, 12 for K, and 0.1 for Mg. It appears clear that organisms were able to develop in solutions that had more than sufficient quantities of the major cations in solution, and their subsequent evolution has masked any correlation that might once have existed.

The third major group of chemical constituents consists of the minor, or trace, elements—minor in volume but certainly not in effect. Generally, trace elements are metals that can have a variety of functions in biota. For example, Mn and Ni serve as enzyme activators, and also have structural influences. Zn and Co, as well as Mn, Fe, Ni, and Cu, can substitute for H^+ as acids. This allows several structural units to be combined simultaneously (coordinated) by the metal ion. Substitutions of metals for H allows the organism to exist over a much broader range of pH conditions than can those that exist using H only.

Another function of metal ions utilizes their variable oxidation states. These redox reactions can affect oxygen availability, as does Fe in hemoglobin and Cu in blue hemocyanin (13), or Mo in the nitrogen cycle. Several excellent summaries of the different functions of the various elements are available in the scientific literature, for example, Refs. 7 through 15. The redox function

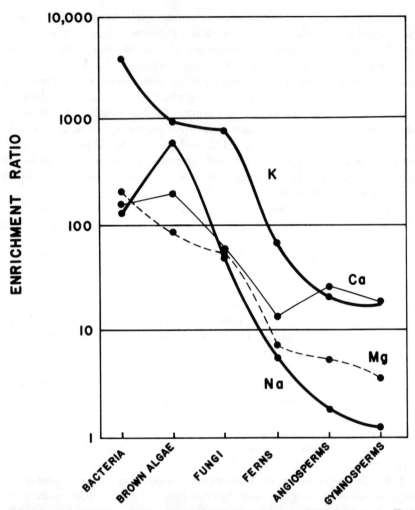

FIGURE 6-4. Major cation compositional variation with biota type. The enrichment ratio is equal to the concentration of the element in a biota type divided by the concentration of manganese in that same type. Sources: Analyses from Refs. 2, 3.

performed by the metallic elements in algae and higher plants was a key process in the geological evolution of the atmosphere and biosphere. In the Precambrian era, blue-green algae and, later, green algae initiated photosynthesis, reducing atmospheric CO_2, and oxidizing Fe and Cu, respectively. Brown algae and ferns in the Paleozoic era concentrated Zn, as terrestrial plants became the dominant form of biota. Today V, Ni, and Mo are concentrated in plant roots, Fe and Mn in leaves, and Cu and Zn in seeds (16). These changes indicate an interrelation between the needs of the plant with the environment (changing oxygen content) and the metallic species present (varying redox potential).

Another aspect of the concentration of elements in specific parts of plants is worth considering. If the proportion of different plant parts varied with time as the different plants evolved, plant fossil accumulations would be expected to show trace-metal variations based on the proportion of plants that make up the assemblage. The variation in trace-metal composition of coals could thus be inherited from the species of coal-forming plants rather than as a result of variations in the process of coal formation. This idea is not new (17, 18) but has been generally ignored in more recent discussions of coal chemistry (19, 20, 21, 22). There are two major reasons why this suggested correlation has not been investigated. First, a complete quantitative identification of the plant species present in each coal is an exceptionally tedious, if not impossible, operation. Second, organic components of coal can gain their trace metals either during growth of the plant precursors or through chemical reactions during the decay process of those plants. As discussed next, the elemental uptake by plants during growth varies tremendously. Perhaps the only conclusion that can be drawn realistically is that extremes in environmental variation will limit the presence of a particular lignin-rich plant, and that this variation (plant—no plant) is the control of trace-element content.

It has been known since the Middle Ages that plants differ in nutritional quality, depending on their geographic location and species. It is now known that the accumulation of elements by plants is the result of a complex interaction of plant properties (species, organ, age of organ, depth of roots, overall health of plant), the soil (chemistry and physical structure, including presence of microorganisms, pH and drainage, availability of element and antagonism of other elements), and the climate (rainfall and temperature). Weather factors (including soil moisture) and plant age are generally dominant over soil characteristics and nutrient supply in determining the content of dry matter and organic constituents in plants (assuming sufficient N, P, K, and Ca). Soil character and nutrient supply are generally the main factors in determining the plants' composition with respect to inorganic constituents, even with the addition of elements by fertilization (23). Weather factors alone are not sufficient to permit description of climate from the biologic standpoint. For example, large variations in weather can result in small plant responses if the variations of all the factors balance out, whereas small weather variations can be catastrophic if the factors vary in an unbalanced way. Unfortunately, the prediction capability of such occurrences is rudimentary at best.

There are three main mechanisms whereby elements can be incorporated into plants: (a) diffusion into the plant from the soil solution; (b) cation exchange at the surface of clay minerals or organic colloids; and (c) absorption involving the aerial parts of the plant. The last mechanism, probably the least significant quantitatively, is an interaction with the atmosphere reservoir rather than with the soil reservoir. Air plants such as Spanish Moss can give a cumulative effect of pollution over time. Plants in the same area but having different ages will have different concentrations of trace element pollutants, indicating variation with time (24).

The presence of an element in the soil solution has, historically, been correlated with the ionic potential (ionic charge/ionic radius) of the element. Elements with a low z/r ($z/r < 2$) value such as Na, K, and Cs occur as soluble cations. Elements with a high z/r ($z/r > 12$) such as S, Mo, and B occur as soluble anionic complexes. If these elements are in solution, they should be more available for incorporation into plant material. Unfortunately, the historical correlation (25) is not obvious in Figure 6-5 where elemental enrichment ratios in plants are shown as a function of ionic potential. It is hard to see that any particular pattern exists. A few of the high ionic potential ions are more concentrated in plants relative to their average composition in soils.

Examination of Table 6-2 indicates that the same pattern holds if the enrichment ratio is based on average crustal composition. Several elements—Zn and Mn, for example—are enriched in plants by many orders of magnitude beyond what would be predicted on the basis of ionic potential alone. Others, such as Ti and Al, fit the pattern predicted. It is probably no accident that Ti and Al are considered biologically inert, whereas Zn and Mn have specific biochemical functions. There are several possible explanations for these discrepancies. First, soil solutions may not be the dominant source of elements for plants; instead, ion exchange between roots and soil particles may be the control. Second, individual variations in plants and associated rocks and soils are too great to make any cumulative plot meaningful. Third, biochemical processes encourage plants to incorporate certain elements regardless of their source (solution, adsorbed, or ion exchange), and also to incorporate elements with similar geochemical behavior—for example, Cd along with Zn. There is insufficient information for one to distinguish among these possibilities; in fact, all of them probably exert an influence.

All of these possible sources of variations must disturb those people who use the elemental ratios to try to demonstrate that life originated elsewhere and was transferred to Earth (30). Enrichment of elements as a function of their ionic potential shows generally similar patterns in sea water, bacteria, plants, and land animals, but this similarity breaks down when all elements are plotted, as in Figure 6-5. It is worthwhile to examine enrichment ratios for these different environments. Figure 6-6 shows a series of enrichment ratios (normalized for Fe concentration ratio = 1) for soil/crust, plant/soil, human/plant, and plasma/human. These diagrams are shown by periodic

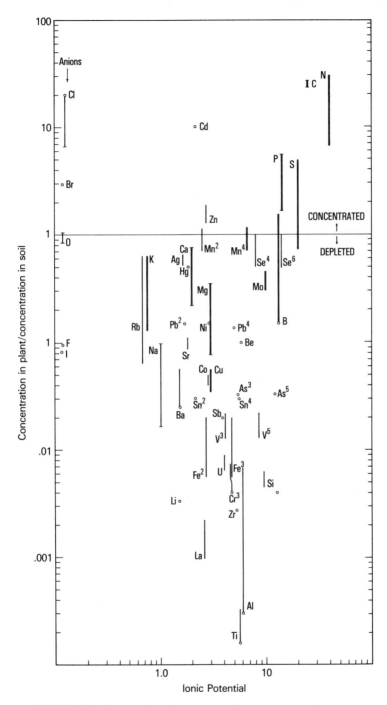

FIGURE 6-5. Plant/soil enrichment ratios of elements as a function of ionic potential. The length of the line indicates the range of analytical values tabulated. Source: Data from Table 6-2.

144

TABLE 6-2
Soil, Crust and Land-Plant Element Compositions

| A | Soil Abundance B | Crustal Abundance C | Land Plants: Dry Weights | | | Enrichment Plant/Soil I(D/B) G | Ratios Plant Soil II(E/B) H |
			D	E	F		
Calcium	24,900	43,000	18,000	5,000	3,778	0.75	.208
Carbon	(B)20,000	170	464,000	(523,000)	393,461	22.7	26.1
Chlorine	(B)100	100	2,000	(660)	496	20.0	6.60
Hydrogen	---	1,000	55,000	(88,000)	65,904	---	---
Magnesium	9,200	24,000	3,200	700	982	.348	.076
Nitrogen	(B)1,000	20	30,000	6,700	5,019	30.0	6.70
Oxygen	(V)490,000	466,000	410,000	(524,000)	524,290	0.837	1.06
Phosphorus	420	1,000	2,300	700	521	5.48	1.67
Potassium	23,000	18,000	14,000	3,000	2,285	.609	.130
Sodium	12,000	23,000	1,200	200	190	.100	.0167
Sulfur	(B)700	300	3,400	500	712	4.86	.714
Arsenic	(B)6	1.9	0.2	---	---	.033	---
Chromium	53	120	0.23	0.4	---	.0043	.0075
Cobalt	10	34	0.5	0.4	---	.050	.040
Copper	25	65	14	9	---	.560	.360
Fluorine	200	600	40–.5	---	---	.100	---
Iodine	5	0.5	0.42	---	---	.084	---
Iron	25,000	57,000	140	500	386	.0056	.020
Manganese	560	900	630	400	211	1.125	.714
Molybdenum	(B)2	1.3	0.9	0.65	---	.450	.325
Nickel	20	95	3	3	---	.150	.150
Selenium	(B)0.2	0.1	0.2	0.1	---	1.00	.500
Silicon	330,000	277,000	200–5,000	1,500	1,208	.0061	.0045
Tin	(B)10	1.9	0.3	---	---	.030	---
Vanadium	76	190	1.6	1	---	.021	.013
Zinc	54	87	100	70	---	1.85	1.30

Element	Soil	Crustal				
Aluminum	66,000	81,000	500	20	556	.0076	.00030
Antimony	(B)3	0.2	0.06	—	—	0.20	.054
Barium	554	450	14	30	—	.025	—
Beryllium	1	1.5	0.1	—	—	.100	.147
Boron	34	7	50	5	—	1.47	—
Bromine	5	2.0	15	—	—	3.00	—
Cadmium	(B)0.06	0.19	0.6	—	—	10.0	—
Germanium	(B)1	1.3	—	—	—	—	—
Lead	20	9	2.7	0.1	—	.135	.0033
Lithium	(B)30	20	0.1	—	—	.0033	—
Mercury	(B)0.03	0.046	0.015	—	—	500	.0083
Radioactive(U)	(V)6	7.3	0.038	0.05	—	.0063	.0010
Rare earth	41	25	0.085	0.40	—	.021	.067
Rubidium	(B)30	90	20	2	—	.67	.50
Silver	(B)0.1	0.09	0.06	0.05	—	.60	.083
Strontium	240	380	26	20	—	.11	—
Tellurium	ND	0.001	2-25	—	—	—	—
Titanium	3,000	6,000	1	2	—	.00033	0.0067
Zirconium	240	130	0.64	—	—	.00027	—

All abundances in ppm.

Source: Soil abundances from Ref. 2 indicated by (B), from Ref. 26, and from Ref. 27 indicated by (V). Crustal abundances from Ref. 28. Land plant abundances, column D from Ref. 2, column E from Ref. 29, column F from Ref. 7.

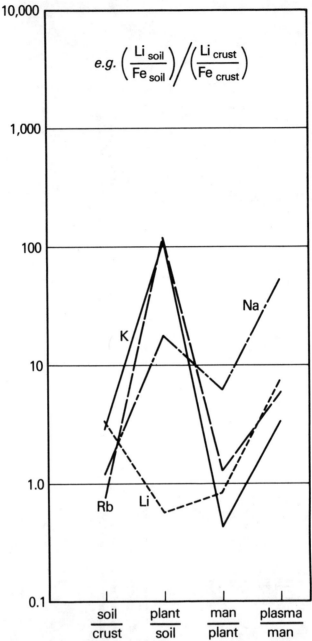

FIGURE 6-6. **Changes in normalized enrichment ratios for selected elements (Fe = 1).** Sources: Crust analyses from Ref. 28; soil analyses from Table 6-2; plant analyses from Ref. 2; man analyses from Refs. 29 and 31; plasma analyses from Ref. 2.

IA - Li, Na, K, Rb

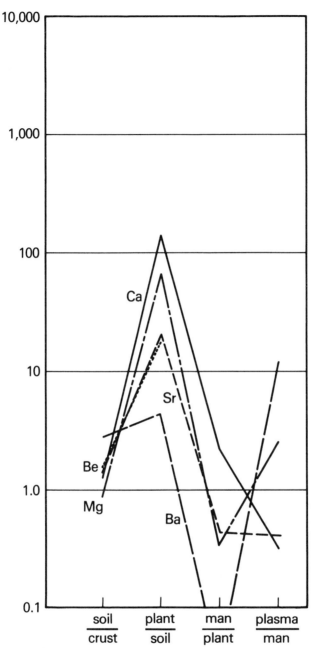

FIGURE 6-6. Changes in normalized enrichment ratios for selected elements (Fe = 1).

IIA - Be, Mg, Ca, Sr, Ba

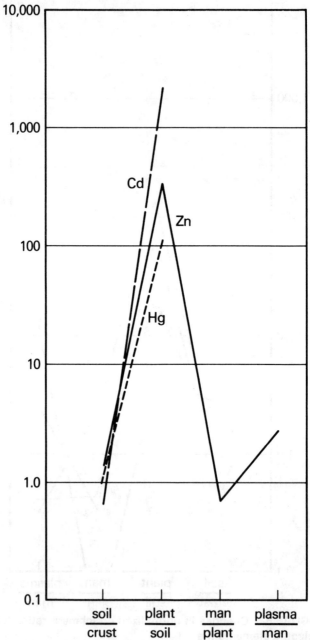

FIGURE 6-6. Changes in normalized enrichment ratios for selected elements (Fe = 1).

IIB - Zn, Cd, Hg

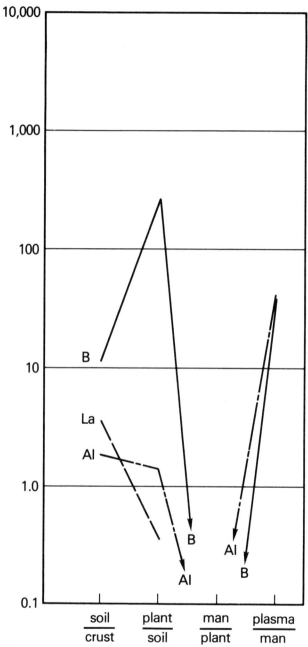

FIGURE 6-6. Changes in normalized enrichment ratios for selected elements (Fe = 1).

IIIA · B, Al, La

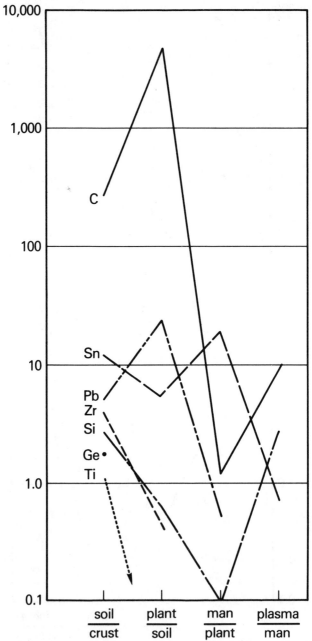

FIGURE 6-6. Changes in normalized enrichment ratios for selected elements (Fe = 1).

IV · C, Si, Ge, Sn, Pb, Ti, Zr

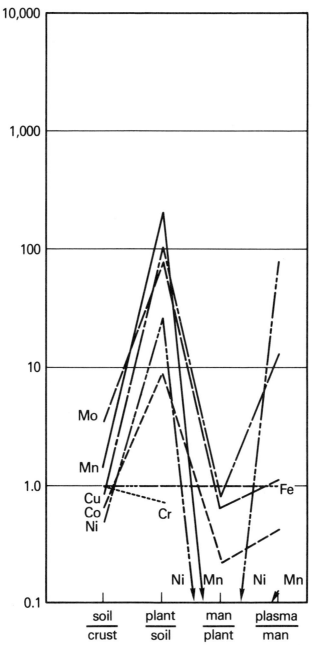

FIGURE 6-6. Changes in normalized enrichment ratios for selected elements (Fe = 1).

Metals · Cr, Mn, Fe, Co, Ni, Cu, Mo

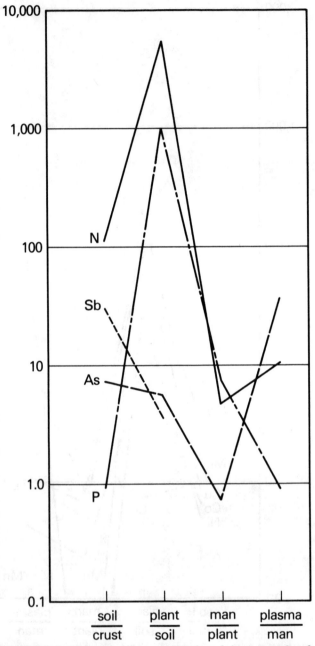

FIGURE 6-6. Changes in normalized enrichment ratios for selected elements (Fe = 1).

VB · N, P, As, Sb

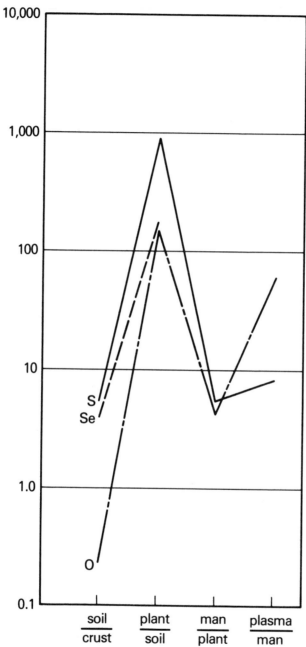

FIGURE 6-6. Changes in normalized enrichment ratios for selected elements (Fe = 1).

VIB · O, S, Se

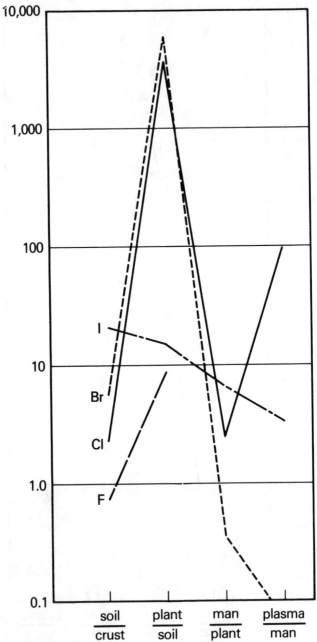

FIGURE 6-6. Changes in normalized enrichment ratios for selected elements (Fe = 1).

VIIB · F, Cl, Br, I

class of the elements. The effectiveness of plants as a factor in concentrating elements (relative to Fe) common in all periodic classes is startling — with the plant/soil value high relative to the soil/crust ratio in virtually all elements. The only elements with a lower plant/soil ratio than a soil/crust ratio include the most inert ones as well as those that have no known biological function, such as Sb, La, Li, Si, Al, Th, As, Sn, Ti, and Zr. The heavy concentration of Cd in plants indicated in Figures 6-5 and 6-6 may mirror the difficulty that plants have in passing Cd once it is in their system, as well as the effectiveness of humans in making Cd available well above its soil/crust ratio.

Whatever the control mechanism, we know that certain plants grow only where or when a particular element is available (*indicator* plants), and other plants tend to concentrate particular elements when possible (*accumulator* plants). Biogeochemistry and geobotany take advantage of these traits. Biogeochemistry depends on the chemical analysis of plants or humus to obtain evidence of "hidden" mineralization. Geobotany involves a visual survey of the vegetation cover to detect mineralization in underlying rocks and soils by using plant distribution or accumulator plants. The major use of biogeochemistry and geobotany in mineral exploration (as indicated by papers published through 1975) had been made by the Soviet Union (50 percent), followed distantly by the United States (17 percent). During the 1965–1975 decade, Australia, England, and India accounted for more than 30 percent of the publications (29, 32, 33). The importance of these techniques in geochemical explorations is well known (29, 34).

Geochemical behavior of the biosphere and its relation to the reservoirs shown in Figure 6-1 can be summarized as follows. First, carbon is absorbed from the atmosphere by the assimilating surfaces of the plant (leaves, shoots, and bark). Concomitantly, nitrogen, mineral elements, and water are removed from the soil reservoir by the root system, and incorporated into the plant. When above-ground parts of the plant die and fall to the ground as litter, subsequent chemical breakdown can return the elements to the soil reservoir — the biogeochemical cycle of Goldschmidt (17). Second, parts of the plants, such as pollen, sap, or leaves, can be removed to provide feed for animals. The elements are incorporated into new organic compounds during the feeding process, and are subsequently returned to the soil from excreta, secreta, or decomposing carcasses of animals. Third, gas exchange takes place between the assimilating surface of the plants and atmospheric air, and between the root surfaces and soil air. The volume of volatile elements transferred through this flux can be quite large, especially for organic compounds. Fourth, plant organs above the ground surface secrete elements during the life of the plant. Stemflow of water can move these secretions back into the soil and water reservoirs quickly.

To understand the full geochemical cycle of the biosphere, we must consider all organisms, such as plants, animals, microflora, microfauna, and bacteria. We have good knowledge of the behavior of the mineral elements and nitrogen in higher plants in the temperate zone, and we have some knowledge of their geochemical behavior in tundras and in the tropics. We know already that the geochemical behavior of elements in the biosphere varies greatly with climate: water behavior varies widely, soil development varies widely—so should not biogeochemistry vary as well? Certain relationships are known and it is those that we will examine next.

Productivity and Biomass

Primary productivity is the amount of biota produced in a given period of time, and is considered to be the internal flux of the reservoirs of the biosphere shown in Figure 6-1. It is a measure of the rate of photosynthetic action and is usually measured in grams per square meter per year ($g/m^2/yr$). *Biomass*, on the other hand, is the reservoir of biota, and it is generally defined as the amount of dry organic matter present at any one time. Biomass usually includes only live organic material, with dead organic material considered as litter or, when decomposed, as soil-organic material. In order to understand the cycling of the elements, values of all should be known. Unfortunately, there is a tremendous amount of uncertainty in the values presently available.

Biomes

Productivity and biomass of a particular area are dependent on the type of biota that exist there, which in turn is controlled by environmental factors such as moisture, temperature, sunlight intensity, seasonal change, and nutrient availability. One recent compilation (35) divided the world into 193 biogeographical provinces each of which fell into one of 14 standard biome types. Another generalized classification (36, 37) of biome type with respect to the environmental conditions of temperature and precipitation is given in Figure 6-7.

Those who grow crops are aware that what can be grown well can vary greatly over small distances with variations in topography, soil type, and water availability. This makes small-scale, local analysis quite different from overall estimates or combinations of analyses. Weighting of measurements is probably the major source of uncertainty about measurements of productivity and biomass. World distribution of the generalized biomes thus has limited value if one is examining the detailed geochemical behavior of a particular region but, as indicated in Figure 6-8, broad groupings are possible. Table 6-3 lists the major biomes and the land-surface areas assigned to each. These areas are constantly changing as forests are cut and cultivated, as desert encroaches

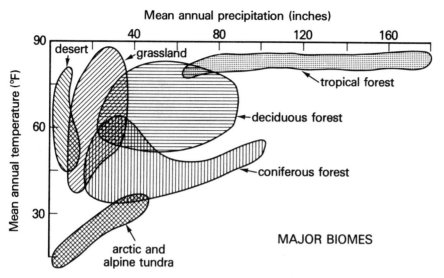

FIGURE 6-7. Distribution of major biomes as a function of temperature and precipitation. Source: Ref. 36.

into grassland, and as irrigation expands cropland. Some controversy exists with respect to the recent value of 10 million km² for the area of tropical rain forests (38) as compared with the older value of 17 million km² (39). Is this discrepancy an indication of the amount of tropical rain forest that has been cut in a 15-year period? Or is it an indication of the increased accuracy of measurement due to the development of remote sensing? Undoubtedly, the answer is both of these, and as the next sections will indicate, such changes of biome type can drastically alter the geochemical behavior of elements.

Productivity

Many studies have been undertaken to correlate the productivity of an area with relatively easy-to-measure environmental variables such as rainfall, temperature, length of vegetation period, and runoff (for example, 37). The environmental variable that comes closest to agreement with present knowledge of productivity patterns is evapotranspiration (42). This stresses again the importance of the water cycle in controlling the geochemical behavior of elements. Table 6-3 clearly shows this effect. Compare the productivity of tropical rain forests (2.3 kg/m²/yr) with that of tundra (0.22 kg/m²/yr). Tundra is generally found in high latitudes where the potential evapotranspiration is very low, whereas tropical rain forests have both a high

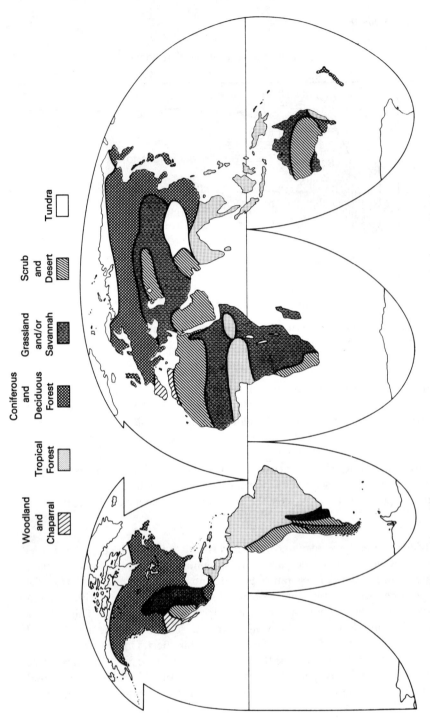

FIGURE 6-8. Distribution of generalized biome types. Generalized from many sources.

potential evapotranspiration and adequate rainfall. This latter condition is not met in extremely dry deserts, for example, and there the net primary production is exceptionally low (0.016 kg/m²/yr).

In order to calculate the total productivity for a particular biome, one multiplies the average value per unit area by the total area. It is now obvious why differences in the assignment of area values can lead to major differences in total productivity values. The present continental total of 133×10^9 t/yr compares to older values of 117.5×10^9 (41), 102×10^9 (40), 121.7×10^9 (42), 100.2×10^9 (43), and 170×10^9 (44). Better weighting of geographical distribution will undoubtably further refine this value.

Note the importance of the productivity of grasslands; grassland plus cultivated annuals account for 47 percent of total productivity but they occupy only one-third of the total land surface. It can also be seen that, on the average, one would expect a decrease in total productivity with each unit-area change from tropical and temperate forest to cropland.

In any steady-state system, the amount of litterfall (leaves, stems, and branches) should equal the new primary productivity. Differences can be accounted for in several ways. First, the difference between net primary production of a biome and its yearly litterfall could represent a yearly increase in the biomass of that biome. Second, we know that that value is a maximum one because crops are harvested, grasslands provide feed, and forests are timbered. Thus, the first gives an indication of the replacement rate of the biomass in the biome, and the second gives the human effect on such replacement. Clearly, it makes a big difference geochemically if the productivity products are added to the biomass in a coral reef or arc or removed from it. Where humans are active there is no such thing as natural geochemistry.

Estimates of marine productivity indicated in Table 6-3 appear to be fairly consistent (39, 45, 46). In the ocean, production is controlled by available sunlight and nutrients. Sunlight availability limits growth in surface water in polar regions, at depth in the open ocean, and in turbid water anywhere. Under other conditions, production is controlled by nutrient availability (a function of local element recycling) and by ocean-water circulation which controls changes in local concentrations (see Chapter 5). It is interesting to note that the disparity between land and sea is exactly the opposite of what is expected (41). If temperature and light are equivalent on both land and ocean, and if water and nutrient availability are the controls, then the land (which has much desert) should be far less productive than the ocean (47). However, as Table 6-3 illustrates, the productivity of the open ocean is comparable to that of desert scrub and tundra and is less than 10 percent that of forests. Clearly, the mobility of nutrients in the sea (primarily by downward settling of organisms and their remains from the zone of photosynthesis) is the limiting factor. There really is no way for short-lived oceanic biota to retain and reuse

TABLE 6-3
Net Primary Productivity, Biomass, and Biomass Distribution of Major Vegetation Units of the World

Biome	Surface Area 10^6 km^2	Net Primary Production kg/m^2/yr	Net Primary Production Total 10^9 t/yr	Litterfall 10^9 t/yr	Biomass Dry Matter kg/m^2	Biomass Total 10^9 t	Biomass Distribution 10^9 t Green Parts	Biomass Distribution 10^9 t Stems Branches, Trunks	Biomass Distribution 10^9 t Roots	Litter Total 10^9 t	Soil Organic Total 10^9 t
Forests											
Tropical rain forest	10	2.3	23	18.5	42	420	38	307	75	6.5	141
Tropical seasonal	4.5	1.6	7.2	5.9	25	112.5	11	79	22	3.8	69
Temperate deciduous	3	1.3	3.9	5.1	28	84	2	69	13	⎱ 18	⎱ 129
Mixed woodland	3	1.5	4.5		30	90	3	75	12		
Boreal	9	0.8	7.2	5.3	22.8	205	10	150	45	31.5	238
Other forest	1.5	1.7	2.6	1.3	20	30	1	25	4	0.8	⎱ 124
Woodland/shrubland	4.5	1.1	5	4.9	11.9	53.5	2	41.5	10	6.3	
Grassland											
Tropical savannah	22.5	1.7	39.3	31.3	6.5	145.7	16	64	65	7.9	455
Temperate	12.5	0.78	9.8	8.6	1.6	20.3	1.8	neg.	18.5	4.9	508
Dwarf and Scrub											
Tundra	9.5	0.22	2.1	1.43	1.4	13.0	2.6	1.3	9.1	23.9	209
Desert scrub	21	0.14	3.0	2.6	0.79	16.5	0.5	1.5	14.5	2.1	290
Extreme Desert											
Dry	9	0.016	0.14	0.14	0.09	0.78	.16	.08	.54	0.14	39
Ice	15.5	0	0	0	0	0	0	0	0	0	0
Cultivated Lands											
Annuals	15	0.9	13.5	6.8	0.07	1.1	.1	0.5	0.5	0.8	221
Perennials	1	1.6	1.6	0.2	5.5	5.5	.5	2.5	2.5		

Other											
Swamps and marsh	2	3.6	7.3	2.1	13.1	26.3	1.6	11.3	13.4	8.8	103
Lakes and streams	2	0.04	0.8	-	0.02	0.04	-	-	-	-	-
Human areas	2	0.2	0.4	0.36	1.6	3.2	0.32	1.4	1.4	0.4	17
Other	1.8	1.0	1.7	-	9.5	16.5	1.0	7.1	8.4	3.0	272
LAND TOTAL	149.3	0.89	133.0	94.7	3.75	1243.9	92 (7%)	837 (68%)	315 (25%)	118.8	2815
Marine											
Open ocean	331.7	.125	41.5		.003	1.0					
Upwelling zones	.4	.5	.2		.02	.008					
Continental shelf	26.6	.36	9.6		.001	.27					
Algae beds/reefs	.6	2.5	1.6		2.0	1.2					
Estuaries	1.4	1.5	2.1		1.0	1.4					
MARINE TOTAL	361.0	.155	55.0	n.a.	.01	3.9	n.a.	n.a.	n.a.		
WORLD TOTAL	510.0	0.369	188.0		2.45	1247.8					

Source: Refs. 38, 39, respectively, for continental and marine productivity, biomass, litterfall, litter and soil organic total. Ref. 40, p.245 for biomass distribution percentages. Woodland and shrubland are estimated from Ref. 40 and assume similar productivities to values from Ref. 6. Soil organic matter assumes that 58 percent of soil organic matter is composed of carbon. Biomass distribution for human areas is calculated at same percent as for annuals. For "other" category - mostly bog - the distribution is the same percent as for swamps. Refs. 39, 41 (see also 38) have estimated animal biomass and productivity. The biomass of animals is 2.01 x 10^9 t, equally divided between oceans and land. The productivity for land animals is 0.83 x 10^9 t/yr for marine animals. n.a. means not applicable.

nutrients, whereas forest has developed on a fixed land surface with large amounts of nutrients stored in the biomass. In addition, cycles on land involving litter, soil, water, and biota operate to suppress nutrient loss. The firm foundation of the land surface allows a higher biomass as well as a higher productivity. Highly productive areas of the ocean are regions where nutrient replacement, reasonable temperatures, and sunlight are all available—but they constitute less than 1 percent of the ocean area. Indeed, as Horne (6, p. 419) pointed out, "the hope that aquaculture will supply man's growing food needs is about as realistic as the hope that magnetohydrodynamics will supply his growing energy needs."

Biomass

Biomass differences between land and ocean areas are even more pronounced than productivity differences, being as much as ten thousand times greater in forests than in the open ocean. As stressed by Whittaker and Likens (39, p. 309),

> stability of surface is again critical. Given stable surfaces, land plants have so evolved that long-lived plants are dominant. These plants use the biomass that is the accumulated profit of net productivity for their extensive root and above ground structures. These structures are in turn part of the basis of high productivity through their support of photosynthetic surfaces and contribution to the pattern of nutrient use and retention. . . . [Plankton are based on] rapid overturns of limited resources with little capital and long term tangible assets.

Knowledge of total biomass is still insufficient to enable us to completely understand the chemical cycles of the elements. It is necessary to know the relative amounts and compositions of the three major plant organ groups: the green part, the above-ground part, and the roots. Elements are concentrated in the green parts during net primary production, and are returned to the cycle with the litterfall of these same green parts, now no longer green. Values in Table 6-3 indicate that the worldwide proportions of these three groups are 7 percent, 68 percent, and 25 percent, respectively, and that the proportions vary greatly with vegetation type. This variability is expected because types such as grasses have few above-ground permanent parts, whereas temperate forests have their weight concentrated in trunk and limbs rather than in needles, leaves, and roots.

It is also interesting to note the relationship of litterfall to green parts. For example, in temperate and mixed forests the litterfall is almost identical to the biomass of green parts; the leaves fall each year. The permanent aspect of leaves in a tropical rain forest can be seen by comparing biomass of green parts to tropical litterfall. Because the green parts are the sites of photosynthesis, cutting in tropical areas has a greater effect on total productivity than it does in temperate woodlands. With most of the biomass in stems, branches, and trunks, it is no surprise that most of the biomass is in forests. This fact also explains the greater amount of biomass in lands cultivated in perennials such as orchards, as opposed to grains, as shown in Table 6-3.

Measurements of annual change in biomass and chemical composition of the different organs and tissues comprising that production give an evaluation of the retention of the annual biomass accretion. Losses are measured from the chemical composition of litter components and of rain water both outside and inside the forest. The sum of the two is uptake, the nutrient requirement of the forest. Accurate measurement of biomass retention involves analysis of the annual wood and bark increments of roots, the above-ground portions, and the one-year-old twigs. Dry weight of whole trees shows good correlation with DBH (diameter 1.3 m above ground) for biomass. Regression curves are used to determine wood and bark increments, a procedure which smooths out local variation.

Measurement of material returned involves measuring the tree litter, ground flora, washing and leaching of the canopy, and stem flow. An amount is also added by throughfall. In many instances, elements dissolved in the rain, those from airborne dust deposited on branches and leaves and then washed, those accumulated on leaves through transpiration, and those leached from twigs cannot be distinguished one from the other. Other imports of elements come from weathering of parent rock (which can be as much as 2 mm/yr for diorite under German beech forests) and nitrogen input from microorganisms and from electrical discharges in thunderstorms. Other losses can occur in drainage water or through harvest. The presence of forest reduces runoff, and thereby the erosion, which could increase the loss and movement of elements as suspended particulate matter in the water (48). Both soil and litter can be removed by this runoff. Differences among very similar organisms are striking as is the behavior of the major nutrient elements in losses due to litter, stem flow, and canopy drip. Seasonal variations also occur. Evapotranspiration decreases runoff. The soluble elements are retained in the system as the water moves through the plants. Element loss should therefore be greatest during the nongreen season. When a forest is clearcut or burned over, there is a general and rapid increase in the concentration of elements in the runoff. Timber harvesting can cause increases in nutrient leaching with concomitant increases in ionic concentrations in streams (49). In areas such as the tropics where the litter reservoir is small or nonexistent (Figure 6-9) this loss may be irreversible due to the lack of a replacement source. In this regard it is interesting to note that trees have evolved mechanisms such as fluttery leaves or uneven canopy to increase evapotranspiration.

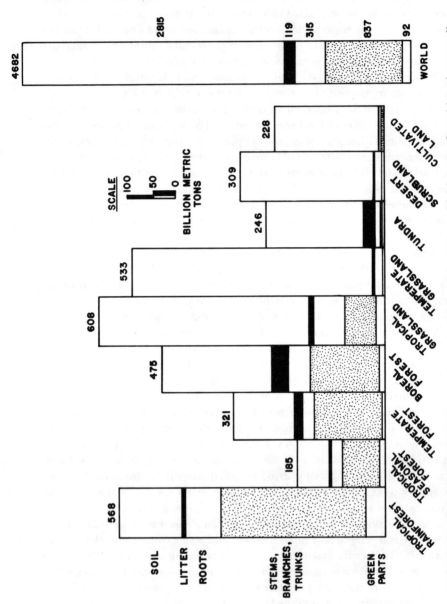

FIGURE 6-9. Distribution of dry organic material in selected world biomes. Considered are soil organic matter; litter; roots; stems, branches, and trunks; and green parts. The world distribution is at a different scale. Source: Information from Table 6-3.

Restitution of the elements to the plant takes place by the removal of these elements from litter. Litter, or dead organic matter, is by definition not part of biota but definitely is part of the biosphere, and has been discussed in Chapter 4. If litter accumulates, the elements may not be recycled and the litter will act as a sink instead of a source. Estimates of the size of the litter reservoir are given in Table 6-3. Accumulation of litter is an indication of a low level of organic decay. Thus the extremes of tundra and tropics are striking, because the litter acts as a reservoir in the tundra, but is quickly broken down and recycled in the tropical warm, moist climates. Thus, litter can act as a reservoir in its own right, or it can be a transient state in the breakdown flux from the biota to the soil organic matter. The relative sizes of the different reservoirs of organic material are indicated in Figure 6-9, which also shows clearly that geochemical behavior differs in the different biomes.

For example, tropical rain forests have the greatest amount of live biomass (roots, stems, branches, trunks, and green parts), boreal forests have the greatest amount of litter, and temperate grasslands have the greatest amount of soil organic matter. Elements that have a tendency to accumulate in one of these reservoirs of the biosphere would thus vary in overall behavior as the total size of the reservoir changes from biome to biome.

An example of annual cycles in a temperate deciduous forest biome is given in Figure 6-10 (50, 51). For a temperate deciduous forest, N, P, and K are concentrated in young leaves rather than in mature leaves; Ca exhibits the reverse concentration. Tree sap has springtime peaks of K, Ca, and N, and a fall peak for K. Also, the same species behaves differently on different soils, illustrating the dominance of soil over climate in temperate climates.

An excellent study of element cycling in a tropical forest (52) made the following key points. Actively growing tissues such as seeds and leaves concentrate nutrient elements more than do less-active tissues such as wood. Because most of the yearly litterfall comes from these rapidly growing tissues, the return of nutrient elements from vegetation to the soil is relatively quick. When this flux is compared to the concentration of elements in the soil, four different groups of elements can be distinguished (52):

1. Those with small soil inventories but rapid recycling rates, where the amount recycled is a significant amount of the inventory: e.g., P, 55%; Fe, 19%; Co, 17%.
2. Those with small soil inventories and slow recycling rates: e.g., Cu, 0.1%; Mn, 0.1%; and Sr, 1.4%.
3. Those with large soil inventories and rapid recycling rates: e.g., K, 56%.
4. Those with large inventories but slow recycling rates: e.g., Ca, 2%; Mg, 2%; and Na, 3%.

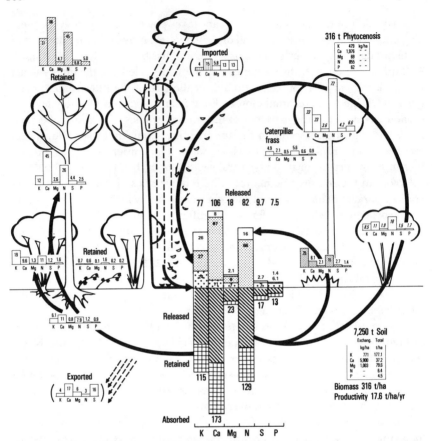

FIGURE 6-10. Annual biogeochemical cycle of elements in a 117-year-old oak-hazel forest in Belgium. *Retained* (cross-hatch) is in increment of trees and shrubs and in perennial parts of herbaceous layer. Total *released* (stripped) is made up of dead aerial parts of herbaceous layer (clear), by tree and shrub litter (light stippled), and by throughfall and stemflow (heavy stipple). *Absorbed* = retained and released. *Imported* by rainfall. *Exported* by drainage water. Source: Ref. 50.

Thus, growth may be controlled by the availability of P and K in tropical rain forests. Rapid breakdown of tropical soils apparently provides a sufficient source for these elements and the rapid breakdown of litter assures their quick availability.

Another major study is a report on the Hubbard Brook area, an overview of about 17 years of observations (53) on a temperate mixed hardwood forest. Two conclusions in that report are of special interest here. First, failure to include annual biomass accretion of nutrients results in a weathering estimate that is 50 percent too low. This emphasizes our previous statements comparing net productivity with litterfall in Table 6-3. Second, the distribution pattern of each element is a complex interaction of sedimentary and atmospheric cycles, biological activity, and climatic variations. For example, more Ca is exported from the watershed through stream runoff than enters by precipitation (53, p. 97–98). Soil must therefore break down to provide the Ca. Of the 62.2 kg/ha/yr of Ca taken up, 40.7 returns in litterfall, 6.7 in stemflow, 3.2 in root litter, 3.5 in root exudates, 5.4 in stems and leaves, and 2.7 in roots.

Of the above-ground biomass at Hubbard Brook, Ca and N have the largest stocks of the elements analyzed. Only about 10 percent of the Ca input comes from precipitation and aerosol fallout, but 100 percent of the nitrogen is supplied in this way. Weathering is the most important source of supply for Ca, Mg, Na, and K. An extensive tabulation (53, p. 105–111) compares the behavior of Ca and other nutrient elements measured at Hubbard Brook with previous studies in similar and different biomes. Unfortunately, only net changes are indicated and the dominant elements are impossible to determine. Why have more such studies not been done? Examination of their figure captions will give a clue to the time and effort required.

If the behavior of elements in a particular biome are known, however, then biogeochemical cycles can be predicted on the basis of litter and litter turnover (40, 44, 50).

Turnover time for Ca was calculated by dividing the amount of Ca present in each category by the average annual flux of Ca from that reservoir through the cycle of soil to wood to canopy to litter to soil (54). If soil time is ignored in order to account for Ca-rich soils, there is excellent agreement between increasing latitude and decreasing rates of uptake and return (longer turnover times). The correlation with length of growing season and with the changes in biome type is excellent. This kind of analysis has led to several generalizations (40):

1. $N \approx Ca$ in temperate deciduous forests
2. $N \gg Ca + K$ in tundra (Indeed, nitrogen dominates in peat, bogs, tundra, coniferous forests, and humid subtropical forests.)
3. $Si > N$ in equatorial forests
4. $Si, Ca > N$ in steppes
5. $Cl > Na > N$ in salt deserts

Figure 6-11 indicates the quantities of annual movement for the major nutrient elements in the different biomes. The correlation with litterfall tabulated in Table 6-3 is striking. Comparison of the litterfall with the amount of litter mass gives an indication of the intensity with which elements are cycled — a pattern that decreases with increasing latitude. As of 1977, 118 areas worldwide were studied in detail (55). Thus, the next decade will

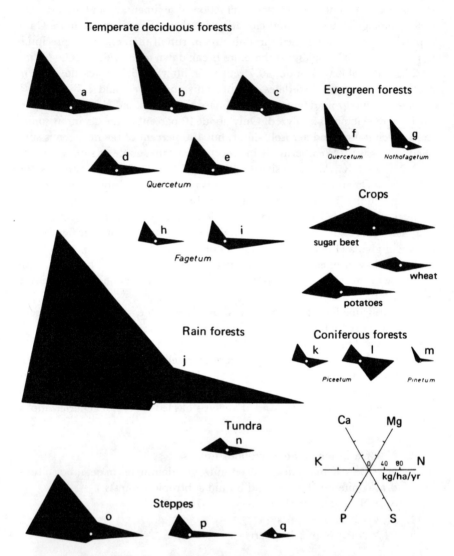

FIGURE 6-11. Comparison of the annual biogeochemical cycle of nutrient elements in different types of terrestrial ecosystems. Source: Ref. 50.

undoubtedly provide some of the detailed information necessary for us to understand better the variations of chemical elements with type of biota — the biogeochemical cycle.

Major Element Cycles

As would be expected from the previous discussions, the behavior of carbon, oxygen, hydrogen, nitrogen, sulfur, and phosphorus is tied completely to the biota reservoir. Because these elements are essential for life, any changes in their cycles are of utmost importance to us. The effect on long-term climatic variations of CO_2 production and accumulation and the effects by combustion on acid rain of SO_2 and NO production have been discussed previously. The complexity of the cycles for the five elements (C, O, H, N and S) is due to their ability to exist in solid, liquid, and vapor forms on the Earth's surface, and to dominate the composition of the atmosphere, biosphere, and hydrosphere. The sixth of these elements, P, may control the limits of the biosphere.

Carbon

There are recent excellent detailed discussions of the geochemical cycle of carbon (56, 57). Several older reviews are also of value (9; 58, esp. p. 368–382; 59; 60, esp. p. 275–277). The inorganic forms of carbon are predominantly CO_2 in the atmosphere, HCO_3^- and CO_3^{2-} in solution, and as CO_3^{2-} carbonate minerals. Through biosynthesis, C is incorporated into organic forms, usually indicated as $(CH_2O)_n$. These organic forms are, in turn, consumed or degraded into inorganic forms. In the atmosphere, the chief gaseous forms of carbon are carbon dioxide (CO_2), methane (CH_4), and carbon monoxide (CO), roughly in the proportions of 300:15:1. Carbon monoxide is produced through combustion and organic breakdown, and is quickly oxidized to CO_2. Methane is produced primarily by decomposition of organic matter in oxygen-poor environments, and is quickly oxidized to CO_2 in the atmosphere.

Total C content in the atmosphere is a sizable reservoir of about 700×10^9 tonnes, comparable to the amount of C present in the biomass, as shown in Figure 6-12. The major fluxes are exchange of CO_2 with ocean water, 100×10^9 t/yr, and respiration, photosynthesis, combustion, and organic decay of continental material. Estimates indicate that the CO_2 content of the atmosphere has increased from 316 to 329 ppm in the last 20 years, and that 5 billion tonnes of C are being added by combustion of fossil fuel each year (62). The effect of the accumulation of CO_2 in the atmosphere is not known, although considerable concern is being expressed about this potential hazard (Refs. 63 through 72).

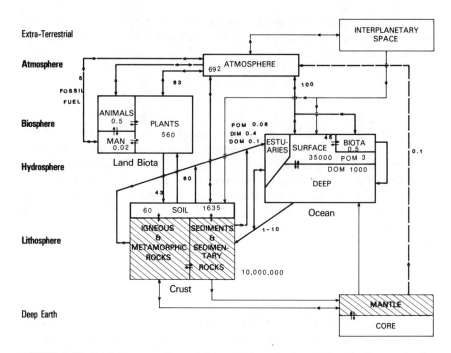

FIGURE 6-12. Carbon reservoirs and fluxes in the geochemical cycle. Reservoirs (inside boxes) are in billion metric tons. Fluxes (along lines) are in billion metric tons/year. See text for abbreviations and references.

The recent eruption of Mount St. Helens has raised questions about the relative effect on the atmosphere of actions by humans and by nature. One recent estimate (60) indicated that the degassing of C from the Earth averages 0.09×10^9 t/yr. This is significantly lower than estimates due to fossil fuel combustion, 5×10^9 t/yr (62).

The carbon cycle in the ocean was discussed in Chapter 5. The major reservoir is as dissolved C, which is present mainly as HCO_3^- (bicarbonate ion). It is many times larger in CO_2-equivalent than the CO_2 available in the atmosphere. The amount of C present in marine biota is small and localized (see Table 6-3) and a significantly smaller reservoir than any of the other oceanic reservoirs. Carbon in dissolved organic material is about 300 times larger than carbon in particulate organic material. Interchange of C between atmosphere and ocean (100×10^9 t/yr) is twice as large in warm-water areas than in cold-water regions (62). The high productivity of ocean photosynthesis is indicated by the value 45×10^9 t/yr. Fluxes into the ocean of dissolved inorganic material (DIM), dissolved organic material (DOM), and particulate organic material (POM) have been discussed in Chapter 3. Precipitation of $CaCO_3$ or settling out of particulate matter is also relatively small, at $1-10 \times 10^9$ t/yr. Effectiveness of the formation and deposition of organic and inorganic $CaCO_3$ with time can be examined by investigating the amount of limestone deposited throughout geologic time, which is well in excess of 10^{15} tonnes. In addition, C in fossil-fuel deposits accounts for another approximate 5×10^{12} t.

The amount of C increases (38) from biota (45 percent) to litter (50 percent) to humus (58 percent). This information has been used with the values of Table 6-3 to determine the value of carbon in the various land reservoirs, resulting in values of 560×10^9 t of biomass, 60×10^9 t of litter, and about 1600×10^9 t of soil carbon. Litterfall, 43×10^9 t C/yr, and productivity, 60×10^9 t C/yr, have been discussed previously. The abundance of terrestrial vegetation has been a major control of atmospheric oxygen for at least the last 200 million years (73, 74).

The stability of soil carbon as a function of climate has been discussed in Chapter 4. The importance of organic material as an adsorbtion substrate, out of all proportion to its quantitative values in streams and sediments, has also been discussed. Residence time for C in the atmosphere is about 4 years, in biota and soils about 40 years, and in ocean sediments and rocks about 400 million years. It is interesting to speculate what would happen if the terrestrial biomass were to decrease by enlarging the oceans. Atmospheric CO_2 would probably increase, as would the temperature. Animals that could not adapt to the changing temperature would die. Forest clearing, fossil-fuel burning, and particulate matter in the lower troposphere may have the same result.

Sulfur

Reservoirs and fluxes for sulfur are indicated in Figure 6-13 (9, 61, 75-83, especially 77, 80 and 82). Even though S exists in all three states and can thus cycle through the atmosphere, the total volume moved is significantly less than that of carbon. Of the 1.8 million metric tons, 0.5 is present as SO_2, added primarily through pollution (6.5×10^6 t S/yr) and secondarily through volcanic emanations (0.3×10^6 t S/yr). The most abundant sulfur compound in the atmosphere is H_2S (1×10^6 t), supplied from biological decay of land plants (5.8×10^6 t S/yr) and marine biota (4.8×10^6 t S/yr). Sulfur is also present as gaseous and particulate sulfates. The most common particulate sulfates are $(NH_4)_2SO_4$ and $CaSO_4$. These particles can be added to the atmosphere through weathering or can form as aersols from sulfates added from sea spray (4.4×10^6 t S/yr) or by the oxidation of SO_2 and H_2S. Removal of sulfur is largest for wet and/or dry deposition onto land (10.6×10^6 t S/yr), onto the ocean (9.6×10^6 t S/yr), or adsorbed by vegetation (1.5×10^6 t S/yr).

The amount of S present in the ocean is 1 billion times larger than the amount in the atmosphere. Practically all of it is present as SO_4^{2-}. Sediments formed can be gypsum ($CaSO_4 \cdot 2H_2O$), anhydrite ($CaSO_4$), or, under

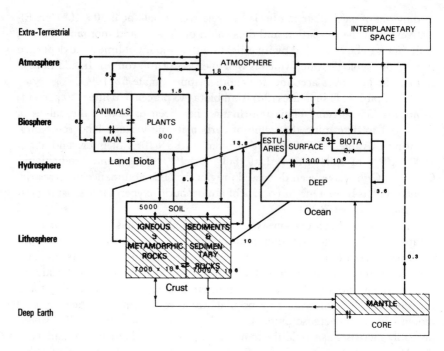

FIGURE 6-13. Sulfur reservoirs and fluxes in the geochemical cycle. Reservoirs are in million metric tons; fluxes are in million metric tons/year. See text for references.

reducing conditions, pyrite (FeS_2). Estimates of the proportions of pyrite and gypsum range from roughly the same to 1:2 (77). This is indicated by the flux of 0.36×10^6 t S/yr as S^{2-} from the breakdown of marine biota out of the 1.00×10^6 t S/yr that precipitates. We do not yet know the importance of sulfur additions from midocean ridges.

Biota take up sulfur primarily as SO_4^{2-}, although some species can reduce SO_2 and others can oxidize FeS_2. The S is needed for protein and, as indicated in Table 6-1, there is a difference in proportion of protein in land and in marine biota. The amount of S present in marine biomass is comparable to that present in the atmosphere. The amount present in land biota is a function of the large biomass on land. Degradation of biota releases sulfur in a reduced form, where, generally, it is oxidized to H_2SO_4—one of the major acids in the soil environment. The amount present in the soil as dead organic matter is about 5000×10^6 t S. Some of this volatilizes into the atmosphere (10.6×10^6 t S/yr), some is buried, and some (8.9×10^6 t S/yr) is eroded into streams where, with weathered rock, it contributes 13.6×10^6 t S/yr to the seas.

Anthropogenic sources of atmospheric S come overwhelmingly from the

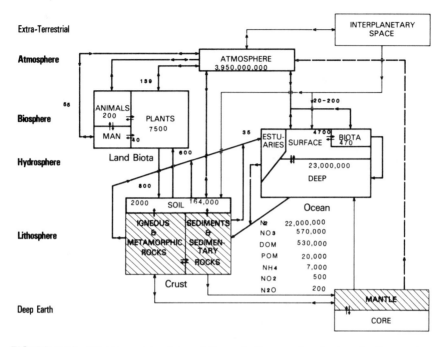

FIGURE 6-14. Nitrogen reservoirs and fluxes in the geochemical cycle. Reservoirs are in the geochemical cycle. Reservoirs are in million metric tons; fluxes are in million metric tons/year. See text for references.

burning of coal. The formation of H_2SO_4 and its effect in acid rain on chemical elements adsorbed on soil and stream sediments have been commented on in Chapter 3.

Nitrogen

The nitrogen cycle has also received some recent major attention (9, 61, 84-90). Nitrogen is unique among the major nutrient elements in that the atmosphere is its largest reservoir (Figure 6-14) with almost 4×10^{15} tonnes. The major species of nitrogen in the atmosphere is N_2—free gaseous nitrogen. There are also about 1300×10^6 tons of N_2O, about 3×10^6 tons of NH_3 and NH_4^+, and about 3×10^6 tons of various other nitrogen oxides (86). The NO is produced through combustion of fossil fuel, and is quickly oxidized to N_2O which reacts with water to form nitric acid. The NH_3 is formed by the decomposition of organic matter, and quickly reacts with water in the atmosphere to form NH_4^+ and OH^-; the NH_4^+ combines with sulfate ions and is removed as a dry precipitate or by rainout.

The ocean is also a sizable reservoir for nitrogen, with most of it present as dissolved N_2. However, the dissolved organic material and the particulate organic material have significant amounts. The concentration of nitrogen in sea water is approximately that found in oceanic biota, indicating a possible equilibrium. The amount of nitrogen in plants, 7500×10^6 tonnes, is calculated from the plant biomass of Table 6-3, assuming a content in plants of 0.6 percent (85). The amount of nitrogen in litter is calculated similarly, using a content in litter of 1.7 percent (86).

The process by which nitrogen is converted from N_2 into biologically usable nitrogen is called *fixing*. The major fixing processes by organisms are by algae and bacteria using photosynthetic energy and respiratory energy to form nitrates—but combustion and industrial processes (to produce fertilizer) are also important. Plants and animals decay to form NH_4^+ compounds, which undergo nitrification or oxidation to nitrates. The total flux of N fixation is about 300×10^6 t/yr, with great uncertainty as to what happens in the oceans (86). The abiological human-caused fixation of 55×10^6 t/yr is a significant fraction of the total nitrogen fixation. Denitrification, or the reduction of nitrates, generally decreases N-availability for plants. Usually, the breakdown of nitrates is carried out by anaerobic bacteria, using the oxygen of the nitrate for respiration. This process was discussed in Chapter 5 for oceanic sediments.

Photosynthesis in marine plants accounts for 4700×10^6 t N/yr (60), with almost all of that recycled. In land biota, photosynthesis accounts for about 600×10^6 t N/yr, and about 800×10^6 t N/yr is added to litter and soil organic material. The remainder is made up by natural fixation and through the addition of fertilizer (40×10^6 t N/yr). The photosynthetic nitrogen flux for marine biota is about 10 times as large as the marine biota nitrogen reservoir. Organic and inorganic runoff to the sea amounts to 35×10^6 t N/yr (88). Fluxes of NH_4^+ among plants, soil, and atmosphere are too complex to be included in Figure 6-14. Indeed, soil scientists and plant scientists know how simplified this diagram already is, but it is important to note the tremendous relative impact that man has on the nitrogen cycle relative to his impact on the cycles of the other major nutrient elements.

Phosphorus

Phosphorus is quite different from the other major elements of the biosphere in that it does not occur in a stable gaseous form. Thus, phosphorus in the atmosphere is either adsorbed on particulate matter from continental sources, or dissolved in sea-spray aerosols. Figure 6-15 indicates the major reservoirs and fluxes for phosphorus (Refs. 91 through 96). The exact amount of P present as aerosols is unknown and the amount removed by wet and dry precipitation

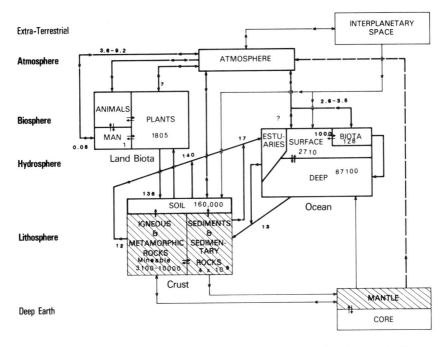

FIGURE 6-15. Phosphorus reservoirs and fluxes in the geochemical cycle. Reservoirs are in million metric tons; fluxes are in million metric tons/year. See text for references.

over land and water is also unknown. The amount released by combustion (0.08×10^6 t P/yr) is only a minor part in comparison with other additions to the atmosphere (95). The soil reservoir is sizable. The amount of P present in continental plants and animals is only about 0.1 percent of the amount present in soils. Plants take P from the soil generally in the form of phosphate $(PO_4)^{3-}$, which in turn is released by the decay of organic matter. The phosphate is not very soluble in water and quickly combines with Ca, Al, or Fe to form insoluble compounds with biologically unavailable phosphorus. These compounds either remain in the soil or are washed into streams and deposited as sediments. As plants pass through their growth cycles, P eventually will have to be added to permit growth. Estimates of the amount of minable phosphorus are about 3 to 10×10^9 tonnes with a flux from mining of about 12×10^6 t P/yr. Unfortunately, much of the minable phosphates will require a large amount of energy for mining and processing.

About 140×10^6 t P/yr are taken from the soil with another 12×10^6 t P/yr added by humans. About 136×10^6 t P/yr is returned to the soil by plant decomposition with about 17×10^6 t P/yr being transported by streams to the ocean.

The P concentration in sea water, like nitrogen, is comparable to its concentration in biota. The flux, about 1000×10^6 t P/yr, from sea water to marine biota, indicates once again the biological churning capacity of marine biota—that is, a given atom of P can be incorporated in the growth and death of 8 different organisms during a year. There appears to be no major source of P into the ocean other than that brought in by river transport, and possibly by particulate fallout from the atmosphere. However, midocean ridge spouts might also be a source for marine P.

Certainly because of the increasing world demand for food, the phosphorus availability and behavior will receive increased attention.

Trace Elements

Trace-element composition of biota is controlled by a variety of factors such as genetic differences, soil and fertilizer effects, climate, and age of organisms (97). Accumulator plants have been discussed previously, with examples of those that concentrate selenium, strontium, aluminum, arsenic, and cobalt. Proportions of plant parts (roots, stems, and leaves) exert control on trace-element distribution, with leguminous plants concentrating cobalt, nickel, iron, copper, and zinc, and with grasses or cereals concentrating manganese, molybdenum, and silicon. The effectiveness with which a plant competes with other biological and inorganic processes for different elements varies according to the plant and the element, but it can be indicated qualitatively merely by the presence of the plant.

If an essential trace element is limited in concentration in the soil, the plant will restrict its growth and/or decrease the concentration of the element in its component parts. Addition of the element artificially causes the reverse. As discussed in Chapter 4, climate and rock character are the key factors in soil development, and they also control trace-element availability. For example, in limestone areas of the eastern United States, soils are rich in phosphorus and potassium, but these soils weather to acid clays that require the addition of lime to be productive. Coastal plain sands, on the other hand, are deficient in iron, manganese, copper, cobalt, and other elements. Shales of the north-central Great Plains are rich in selenium (98). Leaching by downward percolating ground water will also remove trace elements. Water-logged soils can hold trace elements and allow plants to concentrate those elements. Acidity of the soil water often controls the availability of an element by controlling not only its solubility but also its absorption on soil components. Genetic preferences of a plant thus place limits on the pH variation that can be tolerated. As pH increases, the general availability of trace elements decreases with decreasing solubility and increasing adsorption. Addition of lime to the soil increases the pH and, if applied too heavily, can contribute to decreasing trace-element availability. Addition of sulfur to the soil decreases the pH and, as mentioned previously, acid rain alleviates the necessity of adding S.

Seasonal variation in climate can cause variations in trace-element composition of a plant by affecting the soil-water content and the proportions of leaf-stem-seed-root, thus giving variable bulk compositions. As the plant matures, whole-plant concentration of trace elements may increase, decrease, or show no change with stage of growth—depending on the element, plant species, soil, or seasonal conditions (97). Whole-plant concentrations decline with organism age for potassium and phosphorus, and generally for copper, zinc, cobalt, nickel, molybdenum, iron, and manganese; they increase for silicon, aluminum, and chromium. This reflects partly the change in proportion of plant parts, and partly the added storage of inactive elements. If plants are used as a crop, the natural return through litter is circumvented and, eventually, trace elements will have to be returned artificially by fertilization.

The overall effect is that for both plants and the animals that feed upon them, there are generalized patterns of trace-element concentration, as indicated in Figure 6-16. This kind of geographical character gives broad patterns (98, 100, 101, 102) but is not amenable to giving total distribution of trace elements in plants, animals, and humans. Indeed, while recent reviews of trace-element distributions (103, 104, 105) in these biota forms are available, there is serious question whether or not there is actually any *specific* trace-element concentration. One of the complications of plant-animal interaction is that animals whose source of food is from a limited area are more dependent on trace-element concentrations of that soil than those whose food sources cover many different soil types. Animals that eat plants are more dependent on local variation than animals that eat animals. Regional variations in trace elements are common in grazing animals, but rare in humans (although iodine is an obvious exception).

A second factor, the ability of an animal to maintain trace-element concentrations within narrow limits, homeostasis, is most effective for the light metals and least effective for the nonmetals. This balance is maintained by the kidney, liver, and intestines, with adsorption across the gut if the element is needed, and either rejected by the intestine or excreted in the bile or urine if it is not. High exposure can overcome this repulsion mechanism, however, and lead to accumulation (106). Nonmetals that form anions, such as arsenic, antimony, selenium, and tellurium, diffuse quickly across gut walls, react with protein in the red blood cells as sulfur substitutes, and tend to remain in the cell (2). Heavy metals, such as zinc, cadmium, and mercury, have a tendency either to form enzymes that interfere with the above processes, or to substitute preferentially for lighter metals and thus immobilize the enzyme.

A third complication is that humans have introduced trace elements into the environment, both by deliberate use and by unintended pollution. Adequate

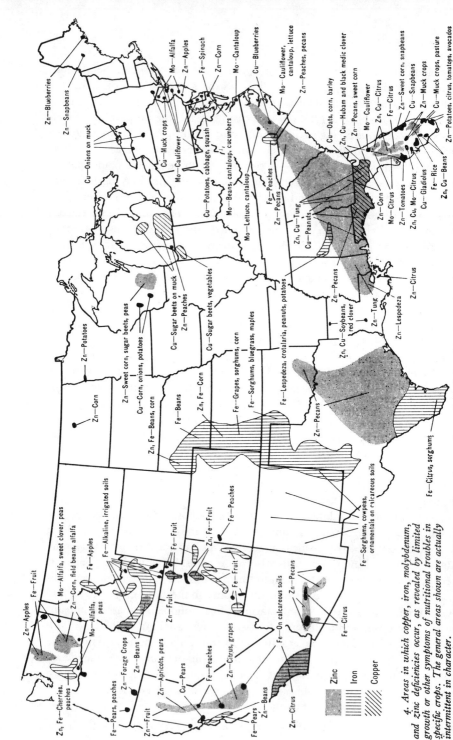

FIGURE 6-16a. Areas in the United States where trace-element deficiencies occur in crops. The areas are intermittent in character. Source: Ref. 99.

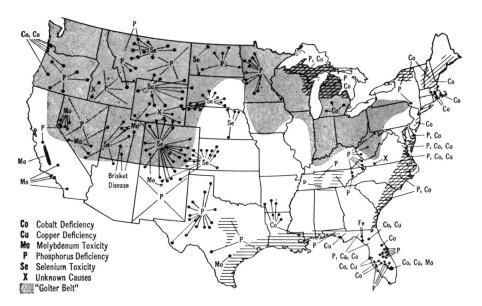

FIGURE 6-16b. Areas in the United States where trace-element–related diseases (toxicity) occur in animals. Shaded area is the Goiter Belt. Source: Ref. 99.

protective mechanisms do not exist because no evolutionary demand has been required. It has been argued that air pollution is more critical to humans than water pollution, because the human organism is better equipped to expel waste material from its digestive tract than from its lungs (107).

The upshot is that in some areas trace elements will accumulate in the biota, whereas in other areas they will not. This leads again to the question mentioned in Chapter 1 as to which are the essential trace elements. The answer apparently depends on the distribution of concentrations (108). The ranges of concentration of both major and trace elements that occur in biota are fairly well known (97, 109, 110).

Health effects of essential trace elements that are present in quantities too low or too high, and of toxic trace elements, continue to receive major attention both in the scientific literature (for example, Refs. 111–116) and through federal support of research. For example, in 1978 at least 13 federal agencies and departments supported research and development work in this area, for a total of almost $600 million (117). A summary of the effects of trace elements on plants and animals is indicated in Table 6-4.

A detailed discussion of trace-element distribution is beyond the scope of this book, but there are several important points that should be discussed. First, trace elements that can exist as a vapor species in the atmosphere have a

TABLE 6-4. Known or suspected effects of anomalous levels of trace elements on plants and animals, including humans. Source: Ref. 112.

Trace Element (Atomic No.)	Environmental Level	On Plants Established	On Plants Conjectured	On Animals Other Than Man[a] Established	On Animals Other Than Man[a] Conjectured	On Man[a] Established	On Man[a] Conjectured
Lithium (3)	low	—	—	—	—	—	Increased incidence of mania
	high	—	Some ability to substitute for K when K is limiting.	—	—	Control mania in some patients.	—
Fluorine (9)	low	—	—	Lower growth rate in rats.[b]	—	Greater incidence of dental caries, especially in young.	Greater incidence of osteoporosis in postmenopausal female; greater incidence of x-ray opaque aortae in male.
	high	Necrosis of tissue with F air pollution.	—	Sparing effect upon Mg-deficient animals re calcinosis; mottled enamel and bone abnormalities, etc. in fluorosis.	—	Mottled enamel. Bone abnormalities.	—
Vanadium (23)	low	Growth stimulation in one species of green algae.	—	Stimulates growth and development in chicks. Lowers serum triglyceride levels.	Deficient O_2 transport pigment in some invertebrates.	—	—
	high	Partial substitute for Mo in NO_3 reductase, but not interpreted as essential.	—	Depresses cholesterol synthesis in rat liver and serum level in young rats. Reduces tuberculous lesions. Toxic; growth depression in rats and chicks.	Affinity for mitochondrial, nuclear fractions of cell. Possible role in bone and enamel mineralization. Negative interaction with Mn re cholesterol synthesis.	—	Lower cholesterol synthesis, possibly age dependent.
Chromium (24)	low	Lower yield and sugar content;[c] but function not determined.	Altered carbohydrate metabolism.	Slower growth rate, impaired glucose tolerance and eye lesions in rats.	Trivalent form in organic molecule with vitamin-like function. Insulin adjuvant.	Impaired glucose tolerance in some human beings responds to Cr^{3+}.	Trivalent form in organic molecule with vitamin-like function. Insulin adjuvant. Progressive depletion in Western society.

Element	Level	Plant effects	Animal effects	Human/clinical	Other
Nickel (28)	high	Toxic; yellow branch disease of citrus and witch's broom of tea.[d]	Experimentally, toxic especially in 6+ valence, but no regional patterns known or expected.	—	—
	low	Sorbed by plants but no evidence of essentiality.	—	—	Ni dust may induce respiratory neoplasia and dermatitis.
Copper (29)	high	Toxic.[d]	Moderately toxic.	—	—
	low	Impaired growth; dieback (exanthema) of citrus shoots;[e] essential.	Regional differences known. Anemia, impaired growth, pigment defects (achromotrichia in cattle, rabbits), connective tissue defects (aneurysm of chicks and pigs), central nervous system amyelination and vacuolization (neonatal ataxia in sheep).	Chick legs show some enlargement of hocks, thickening of bone and abnormal color. Can activate arginase, etc. Concentrates in RNA and possible function there.	Anemia with milk diets. No regional deficiencies known.
Zinc (30)	high	Toxic; may interfere with K metabolism.[e]	250 ppm may improve swine growth yet sheep very susceptible to toxicity; liver damage and hemolytic episodes particularly if Mo low.	Interaction with Zn and Fe.	Wilson's disease patients accumulate Cu due to metabolic "error." Role in aging through increasing free radical formation.[f]
	low	Stunted growth (rosetting), etiolation, poor seed "set" in maize; essential.	Poor growth, skeletal defects, dermatitis, testicular atrophy, altered maternal behavior; exacerbated by Ca and phytate.	Role in insulin production or action.	Hypogonadal dwarfs. Impaired healing and vascular functions including vasospastic disorders. Impaired taste.
Arsenic (33)	high	Toxic at available soil levels beyond approximately 20 ppm.	Interferes with Cu utilization.	—	—
	low	No evidence of essentiality known.	—	Skin and hair appearance less favorable than with As supplement.	—

TABLE 6-4. (continued)

Trace Element (Atomic No.)	Environmental Level	Effects of Anomalous Levels					
		On Plants		On Animals Other Than Man[a]		On Man[a]	
		Established	Conjectured	Established	Conjectured	Established	Conjectured
Arsenic (33) (continued)	high	Plant uptake of As from soils with above average As levels is very limited; but may interfere with phosphorus metabolism.	—	Toxic at high doses, but 5 ppm in drinking water had no adverse effects in lifetime studies with rats and mice. Marine animals tend to have higher As levels than terrestrial animals.	Ameliorates selenosis.	Toxic at high doses.	—
Selenium (34)	low	No evidence of essentiality.	—	Liver necrosis in rats, muscle degeneration in calves, lambs, sterility in rats.	Enhanced by As.	—	Exacerbates kwashiorkor. Possible relation to dental caries, crib deaths,[g] and cancer.
	high	May suppress growth of some plants, but a few species are Se accumulators.	—	Poor growth, sloughed hooves in cattle, deformed chick embryos, etc.	Progressive myocardial failure.	—	—
Molybdenum (42)	low	Poor growth, failure of N fixation by legume root nodule bacteria (NO_3 reductase).	—	Xanthine oxidase cofactor; poor growth; Cu toxicity more likely in sheep.	Impairment of cellulose digestion by rumen flora; increased renal xanthine calculi in sheep.	—	Possible influence on dental caries.
	high	—	—	Weight loss, "teart" scours in cattle. Conditioned Cu deficiency (regional differences partly due to soil drainage and soil reaction rather than Mo level of parent geochemical material alone).	Complex of Mo and Cu.	—	—
Cadmium (48)	low	Absorbed through plant roots but no evidence of essentiality.	—	—	—	—	—

Element	Level	Plants	Animals	Other	Humans	Comments
(cont.)	high					Hypertension. Competes with Zn at metallothionein binding site in kidney. Regional differences in human kidney (Japan and the United States have higher levels).
Tellurium (52)	low	—	—	—	—	—
	high	—	Toxic.	—	—	—
Iodine (53)	low	Not considered essential, yet some plant growth response.[d]	Endemic goiter, stillbirths, hairless newborn pigs. Deficiency may be enhanced by natural goitrogens such as from *Brassica* species and water supplies.	—	Endemic goiter, stunted physical and mental development (cretinism).	Natural goitrogens in some human problems, e.g., walnut in Spain.[h] Interaction with Co.
	high	Toxic (maize)[d] in solution culture.	Experimentally impaired reproduction.	—	Toxic.	Possible problem in one area of Japan.
Lead (82)	low	—	—	—	—	—
	high	Toxic in sand culture; limited translocation from roots during growing season, though air pollution fallout may deposit on above ground portions of plants. Mitochondrial effects in sand culture.[i]	Toxic, shortens life of mice, accumulates in bone.	Impaired protein synthesis. Exacerbates swayback in sheep in low Cu regions.[j]	Toxic, accumulates in bone, central nervous system damage in children, anemia; increased urinary delta aminolevulinic acid is an early sign.	—

[a] Underwood (1971), except as otherwise noted.
[b] Schwarz and Milne (1972).
[c] Bertrand (1967).
[d] Chapman, H. D. [ed] *Diagnostic Criteria for Plants and Soil* (University of California, Berkeley, 1966).
[e] Childers, N. F. [ed] *Minerals Nutrition of Fruit Crops* (Rutgers University Press, New Brunswick, N.J. 1954).
[f] Harman, D., *J. Gerontol.*, 20, 151 (1965).
[g] Money (1970).
[h] Linazasoro, J. M., J. A. Sanchez-Martin, and C. Jimenez-Diaz, *Endocrinology*, 86, 696 (1970).
[i] Miller and Koeppe (1971).
[j] Alloway, B. J., PhD Thesis (University of Wales, Aberystwyth, Sept. 1969).

NOTE: This table was compiled primarily for geochemists by D. J. Horvath at the suggestion of the Workshop participants. The objective is to provide a synopsis, giving probable biological consequences of anomalous levels of those trace elements being considered at the Workshop. Consequently, several biologically important trace elements are not listed. The values "low" and "high" are defined on the basis of biological effects. Absolute levels are not readily available because of the multiple interactions among various elements and the effects of many other variables. An important advantage of its matrix form is that this table highlights areas in which information is seriously deficient or absent.

TABLE 6-5. Composition of U.S. coals. Source: Ref. 22.

Element	All Coal (799 samples)	Anthracite (53 samples)	Bituminous (508 samples)	Sub-bituminous (183 samples)	Lignite (54 samples)	Average Shale
Percent						
Si	2.6	2.7	2.6	2.0	4.9	7.3
Al	1.4	2.0	1.4	1.0	1.6	8.0
Ca	.54	.07	.33	.78	1.2	2.21
Mg	.12	.06	.08	.18	.31	2.55
Na	.06	.05	.04	.10	.21	.96
K	.18	.24	.21	.06	.20	2.66
Fe	1.6	.44	2.2	.52	2.0	4.72
Mn	.01	.002	.01	.006	.015	.085
Ti	.08	.15	.08	.05	.12	.46
Parts per million (ppm)						
As	15	6	25	3	6	13
Cd	1.3	.3	1.6	.2	1.0	.3
Cu	19	27	22	10	20	45
F	74	61	77	63	94	740
Hg	.18	.15	.20	.12	.16	.4
Li	20	33	23	7	19	66
Pb	16	10	22	5	14	20
Sb	1.1	.9	1.4	.7	.7	1.5
Se	4.1	3.5	4.6	1.3	5.3	.6
Th	4.7	5.4	5.0	3.3	6.3	12
U	1.8	1.5	1.9	1.3	2.5	3.7
Zn	39	16	53	19	30	85
B	50	10	50	70	100	100
Ba	150	100	100	300	300	580
Be	2	1.5	2	.7	2	3.0
Co	7	7	7	2	5	19
Cr	15	20	15	7	20	90
Cs	7	7	7	3	7	19
Mo	3	2	3	1.5	2	2.6
Nb	3	3	3	5	5	11
Ni	15	20	20	5	15	68
Sc	3	5	3	2	5	13
Sr	100	100	100	100	300	300
V	20	20	20	15	30	130
Y	10	10	10	5	15	26
Tb	1	1	1	.5	1.5	2.6
Zr	30	50	30	20	50	160

*Average amounts of 36 elements in coal samples and in different ranks of coal, presented on whole-coal basis. For comparison, average amounts in shale are also listed.

much more complicated distribution than those that cannot. Vapor species trace elements, just as the major elements discussed in the previous section, have a much greater opportunity to enter and leave the biogeochemical cycle. These elements, such as mercury, cadmium, selenium, arsenic, antimony, and lead, are generally enriched in the atmosphere relative to their concentration in rocks because of high volatility and/or low boiling point. For water-soluble elements, changes in relative humidity can expedite their transfer from the atmosphere (118). Organic compounds such as methyl mercury often form; they not only exist as a vapor species but are also highly reactive in biological systems. Selenium, tellurium, arsenic, sulfur, tin, platinum, palladium, gold, and thallium generally behave in this way, creating a major concern for the use of Pt in automobile exhaust systems (119).

The second concern is illustrated in Table 6-5, which shows the concentration of trace elements in coal (22). As the United States moves from an oil-dominated energy supply to a coal-dominated one there will be increased concern not only about acid rain but also about the spread of these trace elements.

Certainly many more measurements are needed in order to delineate adequately the natural variations of elements, to measure the species in which they exist, and to document their effects. One difficult aspect is that this problem is generally not experimental science but rather is observational in nature. The process of designing measurements (observations) that give an accurate estimate of the whole is incredibly complex (120). The geochemical atlas of England and Wales (121), a massive undertaking, is one such example. The observations are ambiguous in some cases and do not always support the inferences that are drawn—a circumstance beyond the control of the authors of the atlas. We are still a long way from understanding the natural geochemistry of the trace elements, let alone the perturbations caused by humans.

References Cited

1. Perel'man, A. I., 1967, Geochemistry of Epigenesis; trans. by N. N. Kohanowski, Plenum Press, New York, 266 p.

2. Bowen, H.J.M., 1966, Trace Elements in Biochemistry; Academic Press, London, 241 p.

3. Bowen, H.J.M., 1979, Environmental Chemistry of the Elements, Academic Press, New York, 334 p.

4. Rise Project group: 1980, East Pacific Rise: Hot Springs and Geophysical Experiments; Science, v. 207, p. 1421-1433.

5. Kark, D. M., C. O. Winsen, and H. W. Jannasch, 1980, Deep Sea Primary Production at the Galapagos Hydrothermal Vents; Science, v. 207, p. 1345-1347.

6. Horne, R. A., 1978, The Chemistry of our Environment; J. Wiley and Sons, New York, 869 p.

7. Deevey, E. S., Jr., 1970, Mineral Cycles; Scientific American, v. 223, no. 3, p. 148-158.

8. Speidel, D. H., and A. F. Agnew, 1979, The Natural Geochemistry of Our Environment; p. 77-239 in An Overview of Research in Biogeochemistry and Environmental Health, C.P. 825, Science, Research and Technology Subcommittee, Committee on Science and Technology, U.S. Congress, House of Representatives, Washington, D.C.

9. Garrels, R., F. Mackenzie, and C. Hunt, 1975, Chemical Cycles and the Global Environment: Assessing Human Influences; Wm. Kaufmann, Inc., Los Altos, Calif., 206 p.

10. Kononova, M. M., 1966, Soil Organic Matter, Its Nature, Its Role in Soil Formation and in Soil Fertility; trans. by T. Z. Nowakowski and A.C.D. Newman, Pergamon Press, Oxford.

11. Abelson, P. H., 1978, Organic Matter in the Earth's Crust.; Annual Reviews Earth and Planetary Science, v. 6, p. 325-351.

12. Williams, R.J.P., 1971, Biochemistry of Group IA and IIA Cations; p. 155-173 in Bioinorganic Chemistry, R. Dessy, J. Dillard, and L. Taylor, eds., Advances in Chemistry, v. 100, American Chemical Society, Washington, D.C., 436 p.

13. Hughes, M., 1972, The Inorganic Chemistry of Biological Processes; Wiley, New York, 304 p.

14. Morey, E. R., and D. J. Baylink, 1978, Inhibition of Bone Formation During Space Flight; Science, v. 201, p. 1138-1141.

15. Underwood, E. J., 1977, Trace Elements in Human and Animal Nutrition; 4th ed., Academic Press, New York, 545 p.

16. Boichenko, E. A., G. N. Saenko, and T. M. Udel'nova, 1975, Variation in Metal Ratios During the Evolution of Plants of the Biosphere; p. 507-512 in Recent Contributions to Geochemistry and Analytical Chemistry, A. I. Tugainov, ed., Halstad Press, New York, 695 p.

17. Goldschmidt, V. M., 1937, The Principles of Distribution of Chemical Elements in Minerals and Rocks; Journal of the Chemical Society, 1937, p. 655-673.

18. Leutwein, F., and H. J. Rosler, 1956, Geochemical Investigations of Paleozoic and Mesozoic Coals in Central and Eastern Germany, Freiberg; Forsch., c. 19, p. 1-196 *referenced in* Swaine, D. J., 1975, Trace Elements in Coal; p. 539-550 in Recent Contributions to Geochemistry, A. T. Tugainov, ed., Halstead Press, New York.

19. Zubovic, P., 1966, Physicochemical Properties of Certain Minor Elements as Controlling Factors in Their Distribution in Coal; p. 221-246 in Coal Science, Advances in Chemistry, v. 55, American Chemical Society, Washington, D.C.

20. Gluskoter, H. J., 1975, Mineral Matter and Trace Elements in Coal; p. 1-22 in Trace Elements in Fuel, S. P. Babu, ed., Advances in Chemistry, v. 141, American Chemical Society, Washington, D.C.

21. Gluskoter, H. J., R. Ruch, W. Miller, R. Cahill, G. Dreher, and J. Kuhn, 1977, Trace Elements in Coal: Occurrence and Distribution; Illinois State Geological Survey, Circular 499, Urbana, Ill., 154 p.

22. National Academy of Sciences, 1979, Redistribution of Accessory Elements in Mining and Mineral Processing. Part I. Coal and Oil Shale; Committee on Accessory

Elements, Board on Mineral and Energy Resources, NAS, Washington, D.C., 180 p.

23. Beeson, K. C., and G. Matrone, 1977, The Soil Factor in Nutrition, Animal and Human; Marcel Dekker, Inc., New York, 152 p.

24. Shacklette, H. T., and J. J. Connor, 1973, Airborne Chemical Elements in Spanish Moss; U.S. Geological Survey, Professional Paper 574-E, Washington, D.C., 46 p.

25. Hutchinson, G. E., 1943, The Biogeochemistry of Aluminum and Certain Related Elements; Quarterly Review of Biology, v. 18, p. 1-29, 129-153, 242, 262, 331-363.

26. Shacklette, H. T., J. C. Hamilton, J. G. Boerngen, and J. M. Bowles, 1971, Elemental Composition of Surficial Materials in the Conterminous United States; U.S. Geological Survey Professional Paper 574-D, Washington, D.C., 71 p.

27. Vinogradov, A. P., 1959, The Geochemistry of Rare and Dispersed Chemical Elements in Soils; Consultants Bureau, Inc., New York, 191 p.

28. Beus, A. A., and S. V. Grigorian, 1975, Geochemical Exploration Methods for Mineral Deposits; Applied Publications Ltd., Wilmette, Ill., 287 p.

29. Brooks, R. R., 1972, Geobotany and Biogeochemistry in Mineral Exploration; Harper and Row, New York, 290 p.

30. Barin, A., and J. Navrot, 1975, Origin of Life: Clues from Relations Between Chemical Compositions of Living Organisms and Natural Environments; Science, v. 189, p. 550-551.

31. Mason, B., 1966, Principles of Geochemistry; 3rd ed., J. Wiley & Sons, Inc., New York, 329 p.

32. Hawkes, H. E., ed., 1972, Exploration Geochemistry Bibliography; Association of Exploration Geochemistry Special Volume No. 1, 118 p.

33. Hawkes, H. E., ed., 1976, Exploration Geochemistry Bibliography; Association of Exploration Geochemistry Special Volume No.5, 195 p.

34. Rose, A. W., H. E. Hawkes, and J. S. Webb, 1979, Geochemistry in Mineral Exploration; 2nd ed., Academic Press, New York, 657 p.

35. Udvardy, M.D.F., 1975, A Classification of the Biogeographical Provinces of the World; International Union for Conservation of Nature and Natural Resources (I4CN) Occasional Paper 18, Morges, Switzerland.

36. National Science Foundation, 1975, All That Unplowed Land; Mosaic, v. 6, no. 3, p. 17-21.

37. Lieth, H., 1975, Modeling the Primary Productivity of the World; p. 237-263 *in* Primary Productivity of the Biosphere, H. Lieth and R. Whittaker, eds., Springer-Verlag, New York, 339 p.

38. Ajtay, G. L., P. Ketner, and P. Duvigneaud, 1979, Terrestrial Primary Production and Phytomass; p. 129-180 *in* The Global Carbon Cycle, B. Bolin, E. T. Degens, S. Kempe, and P. Ketner, eds., SCOPE 13, Wiley, New York.

39. Whittaker, R. H., and G. E. Likens, 1975, The Biosphere and Man; p. 305-320 *in* Primary Productivity of the Biosphere, H. Lieth and R. Whittaker, eds., Springer-Verlag, New York, 339 p.

40. Rodin, L. E., and N. I. Bazilevich, 1967, Production and Mineral Cycling in Terrestrial Vegetation; trans. by G. E. Fogg, Oliver and Boyd, London, 288 p.

41. Whittaker, R. H., and G. E. Likens, 1973, Carbon in the Biota; p. 281-302 *in*

Carbon and the Biosphere, G. M. Woodwell and E. V. Pecan, eds., Proceedings 24, Brookhaven Symposium on Biology (CONF-720510), National Technical Information Service, Springfield, Va., 392 p.

42. Lieth, H., 1976, The Use of Correlation Models to Predict Primary Productivity from Precipitation or Evapotranspiration; p. 392-407 *in* Water and Plant Life, Problems and Modern Approaches, O. L. Lange, L. Kappen, and E. D. Schulze, eds., Springer-Verlag, New York, 536 p.

43. Lieth, H., 1975, Primary Production of the Major Vegetation Units of the World; p. 203-215 *in* Primary Productivity of the Biosphere, H. Lieth and R. Whittaker, eds., Springer-Verlag, New York, 339 p.

44. Rodin, L. E., N. I. Bazilevich, and N. N. Rozov, 1975, Productivity of the World's Main Ecosystems; p. 13-26 *in* Productivity of World Ecosystems, U.S. National Committee for the International Biology Program, National Academy of Sciences, Washington, D.C., 166 p.

45. Menzel, D. W., 1974, Primary Productivity, Dissolved and Particulate Organic Matter, and the Sites of Oxidation of Organic Matter; p. 659-678 *in* The Sea, v. 5, Marine Chemistry, E. D. Goldberg, ed., Wiley, New York, 895 p.

46. Fogg, G. E., 1975, Primary Productivity; p. 385-448 *in* Chemical Oceanography, 2nd ed., v. 2, J. P. Riley and G. Skirrow, eds., Academic Press, New York, 647 p.

47. Riley, G. A., 1944, The Carbon Metabolism and Photosynthetic Efficiency of the Earth as a Whole; American Scientist, v. 32, p. 129-134.

48. Pomeroy, L. R., 1970, The Strategy of Mineral Cycling; Annual Reviews of Ecological Systems, v. 1, p. 171-190.

49. Hornbeck, J. W., and G. E. Likens, 1978, Chemical Content of Streams Draining Two Cutover Forests in New England; EOS, v. 59, no. 4, p. 282.

50. Duvigneaud, P., and S. Denaeyer-De Smet, 1975, Mineral Cycling in Terrestrial Ecosystems; p. 133-154 *in* Productivity of World Ecosystems, U.S. National Committee for the International Biology Program, National Academy of Sciences, Washington, D.C., 166 p.

51. Duvigneaud, P., and S. Denaeyer-De Smet, 1970, Biological Cycling of Minerals in Temperate Deciduous Forests; p. 199-205 *in* Analysis of Temperate Forest Ecosystems, D. E. Reichle, ed., Springer-Verlag, Berlin, 304 p.

52. Golley, F. B., J. T. McGinnis, R. G. Clements, G. I. Child, and M. J. Duever, 1975, Mineral Cycling in a Tropical Moist Forest Ecosystem; University of Georgia Press, Athens, Ga., 248 p.

53. Likens, G. E., F. H. Bormann, R. S. Pierce, J. S. Eaton, N. M. Johnson, 1977, Biogeochemistry of a Forested Ecosystem; Springer-Verlag, New York, 146 p.

54. Jordon, C. F., and J. R. Kline, 1972, Mineral Cycling: Some Basic Concepts and Their Application in a Tropical Rain Forest; Annual Reviews of Ecological Systems, v. 3, p. 33-50.

55. diCastri, F., and L. Loope, 1977, Biosphere Reserves: Theory and Practice; Nature and Resources, v. 13, p. 2-7.

56. Bolin, B., E. T. Degens, S. Kempe, and P. Ketner, eds., The Global Carbon Cycle; SCOPE 13, Wiley, New York, 491 p.

57. Golubic, S., W. Krumbein, and J. Schneider, 1979, The Carbon Cycle;

p. 29-46 in Biogeochemical Cycling of Mineral-Forming Elements, P. Trudinger and D. J. Swaine, eds., Elsevier, Amsterdam, 590 p.

58. Woodwell, G. M., and E. V. Pecan, eds., Carbon and the Biosphere; Proceedings 24, Brookhaven Symposium on Biology (CONF-720510), National Technical Information Service, Springfield, Va., 392 p.

59. Woodwell, G. M., R. H. Whittaker, W. A. Reiners, G. E. Likens, C. C. Delwiche, and D. B. Botkin, 1978, The Biota and the World Carbon Budget; Science, v. 199, p. 141-146.

60. Gregor, C. B., 1972, The Carbon Cycle; 24th International Geological Congress, Montreal, Geochemistry, 308 p.

61. Holland, H. D., 1978, The Chemistry of the Atmosphere and Oceans; Wiley, New York, 351 p.

62. Bolin, B., E. T. Degens, P. Devigneaud, and S. Kempe, 1979, The Global Biogeochemical Carbon Cycle; p. 1-56 in The Global Carbon Cycle, B. Bolin, E. T. Degen, S. Kempe, and P. Ketner, eds., SCOPE 13, Wiley, New York, 491 p.

63. Keeling, C. D., 1973, The Carbon Dioxide Cycle: Reservoir Models to Depict the Exchange of Atmospheric Carbon Dioxide with the Oceans and Land Plants; p. 251-329 in Chemistry of the Lower Atmosphere, S. I. Rasool, ed., Plenum Press, New York.

64. Machta, L., 1972, The Role of the Oceans and Biosphere in the Carbon Dioxide Cycle; p. 121-145 in The Changing Chemistry of the Oceans, Nobel Symposium no. 20, D. Dyrssen and D. Jagner, eds., Wiley, New York.

65. Galimov, E. M., 1976, Variations of the Carbon Cycle at Present and in the Geologic Past; p. 3-11 in Environmental Biogeochemistry, v. 1, Carbon, Nitrogen, Phosphorus, Sulfur, and Silenium Cycles, J. O. Nriagu, ed., Ann Arbor Science, Mich.

66. Baes, C. F., Jr., H. E. Goeller, J. S. Olsen, and R. M. Rotly, 1976, The Global Carbon Dioxide Problem; Oak Ridge National Laboratory Publication ORNL-5194, Tenn., 72 p.

67. Baes, C. F., Jr., H. E. Goeller, J. S. Olsen, and R. M. Rotly, 1977, Carbon Dioxide and Climate: The Uncontrolled Experiment; American Scientist, v. 65, p. 310-320.

68. Wallis, J. R., 1977, Climate, Climatic Change, and Water Supply; EOS, v. 58, no. 11, p. 1012-1024.

69. Andersen, N. R., and A. Malahoff, ed., 1977, The Fate of Fossil Fuel CO_2 in the Oceans; Marine Science, v. 6, Plenum Press, New York, 749 p.

70. Woodwell, G. M., 1978, The Carbon Dioxide Question; Scientific American, v. 238, no. 1, p. 34-43.

71. National Academy of Sciences, 1979, Carbon Dioxide and Climate: A Scientific Assessment; Climate Research Board, NAS, Washington, D.C., 35 p.

72. Gribbin, J., 1980, Carbon Dioxide and Climate: The Burning Question; Analog Science Fiction, Science Fact, January, volume C, No. 1, p. 65-73.

73. McLean, D. M., 1978, Land Floras: The Major Late Phanerozoic Atmospheric Carbon Dioxide/Oxygen Control; Science, v. 200, p. 1060-1062.

74. McLean, D. M., 1978, A Terminal Mesozoic "Greenhouse": Lessons From the Past; Science, v. 201, p. 401-406.

75. Hosler, W. T., and I. R. Kaplan, 1966, Isotope Geochemistry of Sedimentary Sulfates; Chemical Geology, v. 1, p. 93-135.

76. Kellogg, W. W., R. C. Cadle, E. R. Allen, A. L. Lazrus, and E. A. Martell, 1972, The Sulfur Cycle; Science, v. 175, p. 587-596.

77. Friend, J. P., 1973, The Global Sulfur Cycle; p. 177-201 in Chemistry of the Lower Atmosphere, S. I. Rasool, ed., Plenum Press, New York.

78. Goldhaber, M. B., and I. R. Kaplan, 1974, The Sulfur Cycle; p. 569-655 in The Sea, v. 5, Marine Chemistry, E. D. Goldberg, ed., Wiley, New York, 895 p.

79. Garrels, R. M., and E. A. Perry, Jr., 1974, Cycling of Carbon, Sulfur, and Oxygen Through Geologic Time; p. 303-356 in The Sea, v. 5, Marine Chemistry, E. D. Goldberg, ed., Wiley, New York, 895 p.

80. Granat, L., H. Rodhe, and R. O. Hallberg, 1976, The Global Sulfur Cycle; p. 89-134 in Nitrogen, Phosphorus, and Sulfur—Global Cycles, SCOPE Report 7, Ecology Bulletin, Stockholm, v. 22, 192 p.

81. Handbook of Geochemistry, 1978, Ch. 16, Sulfur; v. 2, K. H. Wedepohl, ed., Springer-Verlag, New York.

82. Trudinger, P. A., 1979, The Biological Sulfur Cycle; p. 293-314 in Biogeochemical Cycling of Mineral-Forming Elements, P. Trudinger and D. J. Swaine, eds., Elsevier, Amsterdam, 590 p.

83. Krouse, H. R., and R.G.L. McCready, 1979, Biogeochemical Cycling of Sulfur; p. 401-430 in Biogeochemical Cycling of Mineral Forming Elements, P. Trudinger and D. J. Swaine, eds., Elsevier, Amsterdam, 590 p.

84. Paul, E. A., 1976, Nitrogen Cycling in Terrestrial Ecosystems; p. 225-243 in Environmental Geochemistry, v. 1, Carbon, Nitrogen, Phosphorus, Sulfur and Selenium Cycles, J. O. Nriagu, ed., Ann Arbor Science, Mich.

85. Rosswall, T., 1976, The Internal Nitrogen Cycle Between Micro-Organisms, Vegetation and Soil; p. 157-167 in Nitrogen, Phosphorus and Sulfur—Global Cycles, B. H. Svensson and R. Söderlund, eds., SCOPE Report 7, Ecology Bulletin, Stockholm, vol. 22.

86. Södenlund, R., and B. H. Svensson, 1976, The Global Nitrogen Cycle; p. 23-73 in Nitrogen, Phosphorus and Sulfur—Global Cycles, SCOPE Report 7, Ecology Bulletin, Stockholm, v. 22, Svensson, B. H. and R. Söderlund, eds.

87. Bolin, B., and E. Arrhenius, 1977, Nitrogen—An Essential Life Factor and a Growing Environmental Hazard; Report from Nobel Symposium no. 38, AMBIO, v. 6, p. 96-105.

88. Delwiche, C. C., 1977, Energy Relations in the Global Nitrogen Cycle; AMBIO, v. 6, p. 106-111.

89. Simpson, H. J. et al., 1977, Man and the Global Nitrogen Cycle; p. 253-274 in Global Chemical Cycles and Their Alterations by Man, W. Stumm, ed., Dahlem Konferenzen, Berlin, 346 p.

90. Vitousek, P., J. Gosz, C. Grier, J. Meliko, W. Reiners, and R. Todd, 1979, Nitrate Losses from Disturbed Ecosystems; Science, v. 204, p. 469-474.

91. McKelvey, V. E., 1973, Abundance and Distribution of Phosphorus in the Lithosphere; p. 13-31 in Environmental Phosphorus Handbook, E. Griffith, A. Beeton, J. Spencer, and D. Mitchell, eds., Wiley, New York.

92. Gulbrandsen, R., and C. Roberson, 1973, Inorganic Phosphorus in Seawater;

p. 117-140 *in* Environmental Phosphorus Handbook, E. J. Griffith, A. Beeton, J. M. Spencer, and D. T. Mitchell, eds., Wiley, New York.

93. Sawyer, C. N., 1973, Phosphorus and Ecology; p. 633-648 *in* Environmental Phosphorus Handbook, E. J. Griffith, A. Beeton, J. M. Spencer, and D. T. Mitchell, eds., Wiley, New York.

94. Lerman, A., F. T. MacKenzie, and R. M. Garrels, 1975, Modeling of Geochemical Cycles: Phosphorus as an Example; Geological Society of America Memoir 142, p. 205-218.

95. Pierrou, U., 1976, The Global Phosphorus Cycle; p. 75-85 *in* Nitrogen, Phosphorus and Sulfur—A Global Report, B. H. Swensson and R. Söderlund, eds., SCOPE Report 7, Ecology Bulletin, Stockholm, vol. 22.

96. Pierrou, U., 1979, The Phosphorus Cycle: Quantitative Aspects and the Role of Man; p. 205-210 *in* Biogeochemical Cycling of Mineral-Forming Elements, P. A. Trudinger and D. J. Swaine, eds., Elsevier, Amsterdam.

97. Underwood, E. J., 1977, Trace Elements in Human and Animal Nutrition; 4th ed., Academic Press, New York, 545 p.

98. Cannon, H. L., 1969, Trace Element Excesses and Deficiencies in Some Geochemical Provinces of the United States; p. 21-43 *in* Trace Substances, in Environmental Health-III, D. D. Hemphill, ed., University of Missouri, Columbia, 391 p.

99. Beeson, K. C., 1957, Soil Management and Crop Quality; p. 258-267 *in* Soil, U.S. Department of Agriculture, Yearbook 1957, Washington, D.C., 784 p.

100. Cannon, H. L., and W. L. Petrie, 1979, A Review of Recent Activity in the United States; p. 137-150 *in* Environmental Geochemistry and Health, Philosophical Transactions, Royal Society of London, B288, p. 1-216.

101. Connor, J. J., and H. T. Shackletter, 1975, Background Geochemistry of Some Rocks, Soils, Plants, and Vegetables in the Conterminous United States; U.S. Geological Survey Professional Paper 574-F, Washington, D.C., 168 p.

102. Hopps, H. C., 1979, The Geochemical Environment in Relationship to Health and Disease; Interface, v. 8, no. 3, p. 24-29; no. 4, p. 36-38.

103. Mitchell, R. L., and J. C. Burridge, 1979, Trace Elements in Soils and Crops; p. 15-24 *in* Environmental Geochemistry and Health, Philosophical Transactions, Royal Society of London, B288, p. 1-216.

104. Mills, C. F., 1979, Trace Elements in Animals; p. 51-64 *in* Environmental Geochemistry and Health, Philosophical Transactions, Royal Society of London, B288, p. 1-216.

105. Moynahan, E. I., 1979, Trace Elements in Man; p. 65-80 *in* Environmental Geochemistry and Health, Philosophical Transactions, Royal Society of London, B288, p. 1-216.

106. Horvath, D. J., 1976, Trace Elements and Health; p. 319-356 *in* Trace Substances and Health, A Handbook, Part 1, P. M. Newberne, ed., Marcel Dekker, Inc., New York, 398 p.

107. Schroeder, H. A., 1974, The Poisons Around Us, Toxic Metals in Food, Air, and Water; Indiana University Press, Bloomington, 144 p.

108. Liebscher, K., and H. Smith, 1968, Essential and Nonessential Trace Elements; Archives of Environmental Health, v. 17, p. 881-890.

109. Kilmer, V. J., 1979, Minerals and Agriculture; p. 515-558 *in* Biochemical Cycling of Mineral-Forming Elements, P. Trudinger and D. J. Swaine, eds., Elsevier, New York, 612 p.

110. Knisbacher, S., 1979, An Overview of the Biological Activity of the Trace Elements; p. 240-264 *in* An Overview of Research in Biogeochemistry and Environmental Health, Committee Print 825, Committee on Science and Technology, U.S. House of Representatives, 269 p.

111. Gough, L. P., H. T. Shacklette, and A. A. Case, 1979, Element Concentrations Toxic to Plants, Animals and Man; U.S. Geological Survey, Bulletin 1466, Washington, D.C., 80 p.

112. National Academy of Sciences, 1974, Geochemistry and the Environment, v. 1, The Relation of Selected Trace Elements to Health and Disease; NAS, Washington, D.C., 113 p.

113. National Academy of Sciences, 1977, Geochemistry and the Environment, v. 2, The Relation of Other Selected Trace Elements to Health and Disease; NAS, Washington, D.C., 163 p.

114. National Academy of Sciences, 1978, Geochemistry and the Environment, v. 3, Distribution of Trace Elements Related to the Occurrence of Certain Cancers, Cardiovascular Diseases, and Urolithiasis; NAS, Washington, D.C., 200 p.

115. Purves, D., 1977, Trace Element Contamination of the Environment; Elsevier, New York, 260 p.

116. Waldbott, G. L., 1973, Health Effects of Environmental Pollutants; C. V. Mosby, St. Louis, 316 p.

117. McCullough, J. M., A. F. Agnew, D. H. Speidel, and S. Knisbacher, 1979, An Overview of Research in Biogeochemistry and Environmental Health; Committee Print 825, Committee on Science and Technology, U.S. House of Representatives, 269 p.

118. McLaughlin, S. D., and G. E. Taylor, 1981, Relative Humidity: Important Modifier of Pollutant Uptake by Plant; Science, v. 211, p. 167-169.

119. Wood, J. M., 1974, Biological Cycles for Toxic Elements in the Environment; Science, v. 183, p. 1049-1052.

120. Eberhardt, L. L., 1975, Sampling for Radionuclides and Other Trace Substances; p. 199-208 *in* Radioecology and Energy Resources, Ecological Society of America Special Publication No. 1, C. E. Cushing, Jr., ed., Dowden, Hutchinson and Ross, Inc., Stroudsburg, Penna., 401 p.

121. Imperial College of Science and Technology Applied Geochemistry Research Group, 1978, The Wolfson Geochemical Atlas of England and Wales; Oxford Press, 69 p.

7
Geochemical Cycles and Fluxes Revisited: Summary

But at one and the same point of time different breezes go rapidly in different directions.
— Pindar

Water Reservoirs and Fluxes

The earth is a water world, existing at a distance from the sun such that the amount of solar energy received allows water to exist in solid, liquid, and vapor states and to be transformed easily from one state to another. About 2 percent of the water exists as solid in ice caps and glaciers, where its rate of turnover is about 8000 yr. Less than 0.001 percent exists as vapor in the atmosphere, where its rate of turnover is less than 10 days. The remaining water exists in liquid form primarily in the oceans (97.4 percent), with much smaller amounts present as ground water (0.59 percent), lakes and inland seas (0.015 percent), soil water (0.005 percent), river water (0.0001 percent) and water in biota (0.0001 percent). Turnover in the ocean is about 3000 yr for the ocean as a whole, and 300 yr for the surface portions only; for ground water and soil water the turnover rates can vary from a few weeks to many years, and for streams and biota the turnover can range from a few days to several weeks. These turnover times give a measure of the speed and degree with which water enters and leaves a particular reservoir.

These fluxes of water are dominated by the evaporation-precipitation process, with runoff from land to oceans being the difference between precipitation falling on the land and evaporation loss from the land. Inhomogeneity of evaporation and precipitation, caused by the geographic distribution of land relative to oceans coupled with the rotation of the Earth and the intensity of solar radiation, causes the movements of the atmosphere and ocean. For example, behavior of water for the Earth as a whole cannot be counted on to explain what happens in the North Atlantic Ocean or on the South American continent. Continental runoff is concentrated in the Atlantic and Indian

193

oceans, whereas precipitation exceeds evaporation in the Pacific Ocean—making the Pacific a source of water for ocean circulation.

Movement of water in the atmosphere is generally from the equator toward the poles. The amount of water is large—about the equivalent of 20 Amazon rivers across either 40th parallel (1). The ocean-water movement is still not well understood. Generally it sinks in the polar regions as its density increases; then moves southward along edges of the North and South American continents into the deep ocean basins, mixing partially with deep, cold Antarctic waters; and moves eastward through the Indian Ocean and into the Pacific Ocean where it rises and can again reach the atmosphere. This slow circulation takes several thousand years, implying that only the upper layers of the ocean play a role in the much more rapid cycles discussed heretofore.

Transport by Water

Water is also the major means of moving chemical elements from reservoir to reservoir, accounting for about 98 percent of the total volume transferred—with roughly equal amounts of particulate load and dissolved load. The most important control on the type and amount of load is climate, which is the interaction of precipitation and temperature. Downward-percolating water dissolves materials from soils at some levels, and deposits them in others. It provides the necessary water for biota, which are composed of more than 50 percent water. It provides a mechanism for moving the soluble nutrient elements within the soil reservoir and transporting them to streams, estuaries, and oceans.

Soluble compounds such as sulfur dioxide are quickly removed from the atmosphere by rain, not only limiting their upward migration but also strongly affecting the pH of rain and the subsequent stream runoff. Less-soluble materials such as carbon dioxide are much less influenced by water in the atmosphere, and tend to be distributed more evenly. The solubility of materials changes with acidity of the water, generally increasing as the acidity increases. (Rain water and streams are generally acidic, whereas ground water and oceans are slightly basic.) Each interface, such as rain water entering the soil, the B soil horizon, ground water entering streams, streams entering the ocean, or the ocean bottom, is thus a region of chemical change because solubility conditions are quickly altered. Human-caused changes can therefore be very significant if focused on an interface process, such as the production of acid rain, removal of biota affecting soil-water composition, or in a positive sense the mining of manganese nodules from the ocean floor.

The particulate load carried by a stream is much less than the amount of particulate matter produced. Estimates of the proportion of sediment generated to that actually delivered from a watershed range from 50 percent to

5 percent. This material, deposited in stream banks and channels, is "suddenly" made available when the stream discharge drastically increases during floods. Thus, a relatively rare event may be responsible for moving most of the particulate material. It is unfortunate that these kinds of events are the most difficult to measure and predict, and therefore we know very little about them. In any case, estuaries rather than oceans are the repository for stream-carried sediments — primarily because the carrying capacity of the water is a function of its velocity, and that velocity is many times greater in streams than it is in the estuaries. Another cause of particulate deposition in estuaries is the increasing salinity and increasing pH which cause the flocculation of very small particles. In the oceans, organic and inorganic calcium carbonates, siliceous shells and particles, and several types of bottom sediment appear to form in place.

Particles may be more important for what they carry *on* them than what they carry *in* them; that is, adsorption on oxide coatings or organic coatings, or exchangeable ions on clay minerals, can be more significant for some elements than the amount of material carried in solution or in the crystalline particles. Although few studies have been conducted, apparently the chemical elements have a tendency to be transported in all modes simultaneously, the proportion varying with each element. Elements in solution would be transported to the ocean and made available to biota as they process the water; at the other end of the potentially reactive spectrum, elements concentrated in the crystalline form would be least available for interaction with other elements.

Those elements that are adsorbed can cause the most concern. First, they are micronutrient elements, especially the metals, and their availability can control biota growth. Second, they are heavy metals, which are generally toxic to most types of life. Third, adsorption is strongly pH-dependent and is also an interface phenomenon — as indicated by oxide coatings on stream sediments and organic coatings in estuaries. This adsorption is affected by human interaction. Fourth, we really do not know how the elements distribute themselves among the various adsorbing media in the different environments, or why. Elements that are quickly adsorbed and tightly held are as effectively removed from the geochemical cycle as those that are insoluble. Which elements behave this way and under what conditions?

Transport by Air

The 2 percent of material not transported by water is moved by air. Volatile elements follow the general movement of air, but those that are soluble in water are quickly removed. The vertical mixing time within the lower atmosphere is about one month, with interchange between the trade winds and the polar regions taking several months. The doldrums at the equator act as a baffle, limiting the rate of air exchange between the Northern and Southern hemispheres to about a year. In the stratosphere or upper atmosphere, vertical mixing takes about 5–10 yr because of the strong layering, with horizontal movement about the same as in the troposphere but with even less movement between North and South hemispheres (1). The distribution of gases in the atmosphere is strongly influenced by the altitude at which the gases are introduced; they tend to stay in the hemisphere of introduction—an obvious problem when the locations of major industries in the world are considered.

The distribution of particles in the atmosphere is strongly dependent on size. Large particles settle out quickly, whereas smaller ones can be supported by moving air and can be distributed more widely. Major sources of particles are sea-salt aerosols, wind-blown dust, volcanic dust, combustion products, and aerosols forming from atmospheric gases. Sophisticated analyses of sea-surface particles and atmospheric particles have allowed estimates of contributions from natural sources and rough estimates of human contributions. Unfortunately, heavy-element patterns due to fossil-fuel combustion and industrial activity cannot now be distinguished from such patterns produced by volcanic activity.

Predicted increases in Earth temperature as the atmospheric CO_2 increases are expected by some to be balanced by decreased temperature due to atmospheric particles that cause increasing reflection of solar radiation. Particles may increase temperature instead of decreasing it. If so, climatic changes may occur quickly, drastically affecting the patterns of erosion, chemical transport, soil development, and, perhaps most important, biomass and productivity.

The major reservoirs—soil, oceans, and biota—show such tremendous interaction that a "whole Earth" view is the only reasonable one. Rules set for one area do not necessarily apply in another.

The oceans have a much smaller role in the present biosphere than was suspected 20 years ago, because their average productivity is now known to be equivalent to that of a scrub desert. Perhaps it may be that the oceans' major functions are to provide a source of water, to move the water to regions where evaporation demands are high, and to provide the ultimate sink—the ocean sediments.

The distribution of biota is also clearly a function of climate, and the geochemical cycles in each region must be considered individually. That is, chemical elements in a tropical rain forest behave differently than those in a northern spruce forest. If rapid climatic change should occur, it could affect the productivity of the biota, and change the oxygen availability so extremely that all animals would die; or it could affect food availability, whereupon some animals would die. Recent concern about the clear-cutting of tropical rain forests in order to provide farm land has focused on the loss of productivity (the decrease in CO_2 removal per acre of land) and the sudden disruption of the local geochemical cycle without sufficient adaptation time. The role of the oceans as a CO_2 moderator is still uncertain. (See, for example, Ref. 2.)

The distribution of major elements is a function of life itself. The distribution of trace elements is influenced by so many variables that any patterns can be considered quirks of nature—the right rock at the right place at the right time under the right conditions will make a particular element available. In spite of such unpredictability, we must continue our efforts to learn the natural ranges of this variability. Only by so doing can we begin to fully understand the perturbations of man's activities on the natural environment.

References Cited

1. Bolin, B., 1976, Transfer Processes and Time Scales in Biogeochemical Cycles; p. 17-22 *in* Nitrogen, Phosphorus, and Sulfur—Global Cycles, SCOPE Report 7, Ecology Bulletin, Stockholm, v. 22, 192 p.

2. Anderson, N. R., and A. Malahoff, eds., 1977, The Fate of Fossil Fuel CO_2 in the Oceans; Marine Science, v. 6, Plenum Press, New York, 749 p.

Epilogue:
A Moral Dilemma

The morals of a civilization refer to accepted customs of conduct, right living, or correctness of action and attitude in a society. What makes these customs, actions, and attitudes accepted, right, or correct is fundamentally whether they foster the survival of the species. Moral behavior is not something inherited or instinctively practiced by humans, but must be taught to each succeeding generation. The worst individual in a society is considered to be not the one who knows the correct attitude and ignores it (*immoral*), but the one who has never learned (*amoral or without morals*) and has nothing to offer the species. A moral dilemma can be said to occur when no acceptable action is possible or when all of the "correct" actions will have a harmful effect on survival of the species.

Toynbee (1) traced the origin of our present dilemma to the rise and spread of monotheism. In the Greek world, "divinity was inherent in all natural phenomena—in springs and rivers and the sea; in trees, both the wild oak and the cultivated olive tree; in corn and vines; in mountains; in earthquakes and lightning and thunder. . . . Man's greedy impulse to exploit nature used to be held in check by his pious worship of nature" (p. 7). That check was removed by deleting the divinity from everywhere and by concentrating it in one place (or at least a limited number). Indeed, in Genesis God charged: "Be fruitful and multiply, and replenish the Earth, and subdue it; and have dominion over . . . every living thing that moveth upon the Earth." It is not immoral—indeed it is moral in Judeo-Christian belief—to propagate the species, to cut trees for firewood and lumber, to use water and air for carrying away waste, to take metals from the Earth and elements from the soil so as to provide humans with life support.

In this book we have tried to show why these actions cannot be considered independent from one another and what the natural interactions are. The rapid movement of stream water through the hydrologic cycle emphasizes the importance of the material that it transports, keeping in mind that the water cycle is intimately involved in the biosphere. Land that is suitable for farming may not be available without long, slow, and expensive modifications of local geobiochemical cycles. In recent years, the air-land-ocean equilibrium has been disturbed increasingly by humans with consequences that may portend great harm. In other words, we have tried to show why the actions taken by recent civilizations must be questioned as to their "correctness" or moral value. If such actions threaten the species itself, then they must be considered immoral. A dilemma exists!

What to do? Scientists will state that they need more information, that they do not know whether any new action taken will help or harm the situation, and that more information will aid in that decision. This book makes clear that the inhomogeneity of nature makes the gathering of such information difficult, time consuming, and mostly local in value.

Technologists will state that humans have no major worry other than the quick development of new sources of energy, and then these problems will take care of themselves. But technologists often ignore the effect of population; Lvovitch (2) stated (p. 20) "By the year 2000 the quality of water treatment is sure to be improved by two and probably four or five times. By the same year the volume of waste waters will have increased approximately ten to fifteenfold. This suggests that even if all waste waters are treated and if treatment methods become much more refined, the degree of river water pollution will still have increased no less than three times."

Finally, the environmental activists will cry for a halt, for they see an immediate change in moral behavior as the only possible solution to our dilemma. They believe that a return to the religious treatment of the Earth is essential. The complete and immediate withdrawal of humans from nature is as unrealistic an attitude as that of the wanton technologist. The species may survive, but as what?

We humans must learn to accommodate our lives to the geochemical cycles of the Earth. We should use the Earth's resources freely when we know that there is no harm and restrict our use when there is. As Whittaker and Likens (3) stated: "Leadership is called for, in which major nations direct their efforts and those of others that can be influenced toward policies based on longer term accommodation of population and industry to the limitations of the world. If such policies are not possible; if governments are still impelled to maximize their nation's wealth and power, the future may show little of what might have been hoped for, given human intelligence and technologic power, but will be more typical of history."

We agree.

References Cited

1. Toynbee, A. J., 1973, The Genesis of Pollution; Horizon, v. 15, no. 3, p. 4–9.
2. Lvovitch, M. I., 1977, World Water Resources, Present and Future; AMBIO, v. 6, no. 1, p. 13–21.
3. Whittaker, R. H., and G. E. Likens, 1975, The Biosphere and Man; p. 305–328 in Primary Productivity of the Biosphere, H. Lieth and R. Whittaker, eds., Springer-Verlag, New York, 339 p.

Index

Absorption, 62, 122, 125, 142
Acre-feet, 14(table note)
Adsorption, 51, 61, 69, 70, 71, 97,
    106, 107, 108, 110, 126, 127,
    155, 171, 173, 174, 177, 195
Aerosols, 53, 54(table), 59, 61–62,
    108, 110, 167, 171, 174, 196
Africa, 17(table), 23, 25(table), 34,
    85
Ag. *See* Silver
Agriculture, Department of, U.S., 81
A horizon. *See* Soil, surface
Air pollution, 179
Al. *See* Aluminum
Alcohols, 137
Algae, 110, 137, 140
    slimes, 67
Alkaline particles, 62
Al_2O_3. *See* Alumina
Alumina (Al_2O_3), 85, 97
Alumino silicates, 114
Aluminum (Al), 3(table), 41, 55, 61,
    90, 94, 95, 99, 117, 125, 126,
    142, 155, 176, 177
    hydroxides, 84, 97
    oxides, 84, 85, 90
    oxyhydroxide, 84
    and phosphorus, 175
Amazon River, 24, 26–27(table), 40,
    46, 68(figure)
    basin, 34
    estuary, 115
    tributaries, 43
Amines, 137

Amino acids, 110
Ammonia (NH_3), 62, 104, 137, 173
Ammonium (NH_4), 94, 104, 173, 174
Ammonium sulfate ((NH_4)$_2SO_4$), 59,
    62, 66, 171
Ammonium sulfide (NH_4HSO_4), 62
Anemia, 4
Anhydrite ($CaSO_4$), 171
Animals. *See* Biota; Plant, -animal
    interaction; Trace elements, and
    disease
Anions, 69, 70(figure), 97, 106,
    107(table), 127, 138, 177
Antarctica, 17(table), 18, 21, 23,
    25(table), 52, 120
Antarctic Current, 16, 120
Antimony (Sb), 3(table), 61, 155, 177,
    185
Aquaculture, 162
Arctic Ocean, 15, 17(table), 18, 19,
    29, 30(table), 31, 52, 120,
    121(figure)
Argon, 52
Aridosols. *See* Soil, desert
Arizona, 55
Arsenate (AsO_4), 97
Arsenic (As), 3(table), 60, 61, 97, 155,
    176, 177, 180–183(table), 185
Arteries, 139
Arthritis, 5
As. *See* Arsenic
Ash, 62. *See also* Fire; Volcanoes
Asia, 17(table), 23, 25(table), 29, 34,
    45, 66

AsO₄. *See* Arsenate
Aswan dam, 46
Atlantic Ocean, 15, 16, 17(table), 24, 29, 30(table), 31, 61, 111, 112(figure), 117, 118, 120, 121(figure), 126, 127, 193
Atmosphere, 7(figure)
  carbon dioxide content, 169, 196
  and geochemical movement, 52-66, 195-196
  nitrogen in, 173
  oceanic, 58(table), 60-61, 108, 133, 170
  particulate load, 54, 171, 196
  residence time, 61, 62
  and soil interface, 133
  sulfur in, 172
  urban, 58(table), 59
  See also Trace elements, vapor; Water, atmospheric
Atomic particles, 9
Australia, 17(table), 21, 23, 25(table), 34, 40, 155
Authigenic material, 114, 120, 122, 125

B. *See* Boron
Ba. *See* Barium
Bacteria, 137, 140(figure), 156
  anaerobic, 174
Baikal, Lake, 21
Baltic Sea, 15
Barium (Ba), 3(table), 111, 115
Basaltic material, 125, 126
Bases, 90
Bauxites, 85, 90
Beech forests, 163
Belgium, 166(figure)
Benguela Current, 16
Benthic Current, 117
Benthic life forms, 49, 117
Bering Strait, 18
B horizon. *See* Soil, sub-
Bicarbonate (HCO₃), 41(table), 42(figure), 43, 44(table), 51, 64, 83, 99, 106, 113, 114, 115, 138, 169, 170
Biogenic particles, 115, 117, 118, 125, 127, 195
Biogeochemical cycle, 97, 155, 166(figure), 167-176
Biogeochemistry, 155
Biogeographical provinces, 156
Biological action zone, 4(figure), 5
Biomass, 131, 159, 160-161(table), 162-163, 165-169, 171, 172
  defined, 156
Biomes, 156-162
  elements, 167-169
Biosphere, 7(figure), 8, 88, 132(figure), 137, 155-156, 169, 196, 199
  defined, 131
Biota, 131, 169
  defined, 131
  element ratio in, 134, 136(table), 171, 172
  land, 131, 132(figure), 133, 134, 135, 136(table). *See also* Ocean, biota
  primary productivity, 156, 157, 159-162, 167, 196
  trace elements in, 139, 176-185
  water in, 193
Birch trees, 31
Birnessite (δMnO₂), 127
Black Sea, 100, 113
Blood clotting, 139
Blue hemocyanin, 139
Bogs, 91, 167
B(OH)₄⁻. *See* Boron hydroxide anion
Bone brittleness, 5
Bone diseases, 4
Boron (B), 2, 3(table), 114, 127, 142
Boron hydroxide anion (B(OH)₄⁻), 114
Br. *See* Bromine
Brahmaputra River, 24, 26-27(table), 45
Brazil Current, 16
Bromine (Br), 3(table), 103

Ca. *See* Calcium
CaCO$_3$. *See* Calcium carbonate
Cadmium (Cd), 3(table), 61, 67, 69, 107, 111, 125, 142, 155, 177, 180–183(table), 185
Ca-HCO$_3$/Na-Cl ratios. *See* Calcium-carbonate/sodium-chloride ratios
Calcite, 117
Calcium (Ca), 3(table), 39, 41, 43, 44(table), 51, 60, 61, 83, 86, 90, 94, 95, 103, 106, 113, 114, 117, 122, 125, 138, 139, 141, 165, 167
  and phosphorus, 175
Calcium carbonate (CaCO$_3$), 90, 104, 106, 113, 114, 115, 117, 122, 138, 170–171, 195
Calcium-carbonate/sodium-chloride (Ca-HCO$_3$/Na-Cl) ratios, 41
Calcium phosphate, 138
Calcium sulfate (CaSO$_3$), 43, 138
Calcrete, 91
Caliche, 91
California Current, 16
Canadian shield, 52
Canaries Current, 16
Canopy drop, 31, 167
Carbohydrate building block (CH$_2$O)$_n$. *See* Carbohydrates, polymers
Carbohydrates, 110, 134, 137
  polymers (CH$_2$O)$_n$, 2, 134, 135, 169
Carbon, 41, 97, 110, 113, 137, 155, 169
  cycle, 169–171
  residence time, 171
Carbonate anion (CO$_3^-$), 169
Carbonate-compensation depth (CCD), 118–120
Carbonate rocks, 23, 43, 44(table)
Carbonates (CO$_4$), 67, 106, 111, 114, 117–120, 123, 124(table)
Carbon dioxide (CO$_2$), 62, 83, 111, 113, 114, 120, 135, 137, 140, 169, 170, 171, 196, 197
  solubility, 18
Carbonic acid (H$_2$CO$_3$), 99, 113
Carbon monoxide (CO), 169
Carbon/nitrogen ratio (C/N), 134, 135
Carbon/sulfur ratio (C/S), 134
Carboxylic acid, 97
Cartilage, 139
CaSO$_3$. *See* Calcium sulfate
CaSO$_4$. *See* Anhydrite
CaSO$_4 \cdot$ 2H$_2$O. *See* Gypsum
Caspian Sea, 21, 24
Cataracts, 139
Cations, 64, 69, 70(figure), 94(figure), 97, 106, 107(table), 114, 115, 122, 138, 139, 140(figure), 141–142
CCD. *See* Carbonate-compensation depth
Cd. *See* Cadmium
Ce. *See* Cerium
Cells, 134, 135(figure), 138–139
Cellulose, 134, 135, 138
Cereals, 4
Cerium (Ce), 61
Cesium (Cs), 142
Chemical precipitation, 9
Chemical weathering, 9, 81, 100, 114, 167
  reverse, 115
  See also Soil, weathering
Chesapeake Bay, 49, 67
CH$_4$. *See* Methane
Chitin, 135
Chloride, 104
Chlorine (Cl), 2, 3(table), 41(table), 42(figure), 43, 44(table), 51, 55, 59, 60, 83, 103, 138, 167
Chlorite, 84, 99
Chlorosis. *See* Plant, chlorosis
C horizon. *See* Soil, weathered parent rock
Chromate (CrO$_4$), 113
Chromium (Cr), 3(table), 61, 177, 180–183(table)

Chromium hydroxide ($Cr(OH)_2$), 113
($CH_2O)_n$. *See* Carbohydrates, polymers
Cl. *See* Chlorine
Clays, 43, 84, 88, 90, 97, 110, 114, 115, 118, 122, 123, 124(table), 126, 142, 176, 195
Climate
  and sediment, 45, 46(figure)
  and soil formation, 88–91
  See also Erosion, and climate; Vegetation
Clouds, 61, 62
C/N. *See* Carbon/nitrogen ratio
Co. *See* Cobalt
CO. *See* Carbon monoxide
Coal, 131, 135, 141, 173, 184(table), 185
Cobalt (Co), 3(table), 61, 122, 125, 126, 138, 139, 165, 176, 177
Colloids, 64, 66, 67, 90, 97, 126, 142
Colorado River, 46–47
Combustion, 169–170, 173, 175, 196
Congo River, 24, 26–27(table), 39, 45
  basin, 34
Coniferous trees, 90, 157(figure), 158(figure), 160–161(table), 168(figure)
Copper (Cu), 3(table), 4, 5, 60, 61, 67, 104, 111, 122, 125, 126, 127, 139, 140, 141, 165, 176, 177, 180–183(table)
Coriolis effect, 18
CO_2. *See* Carbon dioxide
CO_3^-. *See* Carbonate anion
CO_4. *See* Carbonates
Cr. *See* Chromium
CrO_4. *See* Chromate
$Cr(OH)_2$. *See* Chromium hydroxide
Croplands, 164(figure), 168(figure), 199
  annual and perennial, 160–161(table), 163
Crop management, 99
Crystallization, 67, 138–139, 195.
  See also Precipitation, and salinity

Cs. *See* Cesium
C/S. *See* Carbon/sulfur ratio
Cu. *See* Copper
Cultivated land. *See* Croplands

Darling River, 45
Davies, T. A., 120
Davis Strait, 18
DBH. *See* Diameter at breast height
Deciduous forest, 157(figure), 158(figure), 160–161(table), 165, 166(figure), 167, 168(figure)
Deep earth, 7(figure), 8
Deep seepage, 32
Denitrification, 125, 174
Denmark Strait, 18
Denver (Colo.), 43
Deposition, 9, 14, 24, 29, 117, 195
  dry, 65, 171
  and human health, 139
  See also Sediment
Desert, 157(figure), 158(figure), 159, 160–161(table), 164(figure), 167.
  See also Soil, desert
D horizon. *See* Soil, unweathered rock
Diabases, 85
Diagenesis, 122
Diameter at breast height (DBH), 163
DIM. *See* Dissolved inorganic material
Diorite, 163
Dissolved inorganic material (DIM), 170. *See also* Dissolved load data; Elements; Geochemical ecology, movement
Dissolved load data, 37, 39–41, 45, 46(figure), 120, 194
Dissolved organic material (DOM), 170. *See also* Dissolved load data; Organic matter
Distilled water, 62
Doldrums, 19, 195–196
DOM. *See* Dissolved organic material
Drainage, 23–24, 28(figure), 47, 163
Dunites, 85
Dust, 54(table), 55, 59, 60, 62, 66, 115, 196

Index 205

Earth, 9, 13, 16(table), 17(table), 52.
  See also Deep earth; Elements, of Earth's crust
East African lakes, 21
East Pacific Rise material, 125, 126, 137
Ecumene, 81
Elbe River, 67, 70
Electrolytes, 2, 3(table), 69
Electromagnetic radiation, 9
Elements
  of Earth's crust, 5, 6(figure), 7
  necessary, 2, 3(table), 5
  nonessential, 3(table), 5
  nonvolatile, 137–138
  oceanic sources of, 60. *See also* Ocean, elements in
  solubility, 97–99, 138, 142, 185, 195
  See also Biogeochemical cycle; Biota, element ratio in; Geochemical ecology, movement; Plant, elements; Soil, elements in; Trace elements
Eluviated zone. *See* Soil, surface
England, 67, 155, 185
Enzymes, 139, 177
Erosion
  and climate, 45
  and forests, 163
  ice, 52. *See also* Water, erosion
  prediction, 48–49
  slope, 49
  soil, 53, 55, 56–57(figure), 81, 99–100, 138
  wind, 9, 38(table), 53–54. *See also* Atmosphere, and geochemical movement
Estuarine interface, 24, 29, 49–51, 67, 70, 111, 122, 194, 195
Eu. *See* Europium
Europe, 17(table), 23, 24, 25(table), 29, 34
Europium (Eu), 61
Evaporation, 13, 18, 19, 21, 23, 24, 29, 31, 32, 34, 61, 90, 193, 194
  and salinity, 43, 103

  See also Water, flux
Evapotranspiration, 22, 34, 90, 91, 157, 159, 163
Evolution, 138, 140
Exogenic cycle, 8, 9
Extra-terrestrial, 7(figure), 8
Exudation, 53

F. *See* Fluorine
Fe. *See* Iron
Feldspar, 84, 99, 114
$Fe(OH)_3$. *See* Iron, hydroxide
FeOOH. *See* Goethite
Fermentation, 137
Ferns, 140
Ferric oxide (Fe_2O_3), 85
Ferromagnesian minerals, 43, 84
Ferromanganese particles, 123
Fertilizer, 97, 138, 176, 177
FeS_2. *See* Pyrite
Fe_2O_3. *See* Ferric oxide
Fire, 53, 54(table)
Flocculation, 24, 51, 67, 195
Flooding, 47–48
Flood plains, 22
Fluorine (F), 3(table), 180–183(table)
Flushing velocity, 49, 51
Fluxes, 7, 8, 9, 137
  carbon, 169–171
  See also Erosion, soil; Ocean, fluxes; Water, flux
Fog, 62
Food and Nutrition Board, 5
Forested areas, 31, 62, 90, 157(figure), 158(figure), 159, 160–161(table), 162, 163, 164(figure), 165, 167, 168(figure), 196–197. *See also* Podzols
Fossil-fuel deposits, 122. *See also* Coal; Oil
Fraser River, 24, 26–27(table)
Free nitrogen (N_2), 113, 114, 173, 174
Free oxygen (O_2), 83, 88, 125, 137
Fuego volcano (Guatemala), 55, 59
Fulvic acids, 95, 99, 110

Ganges River, 24, 26-27(table), 45
Garrels, R. M., 20
Gelbstoffe, 108
Geobotany, 155
Geochemical atlas, 185
Geochemical ecology
 cycles, 8, 104(figure), 155-156, 195, 196-197, 199-200. *See also* Biogeochemical cycle
 defined, 2
 movement, 37, 38(table), 52, 113, 134, 167, 168(figure), 197. *See also* Atmosphere, and geochemical movement; Erosion, wind; Rivers, and geochemical movement; Streams, and element transport
 See also Fluxes; Reservoirs
Geological Survey, U.S., 22, 23
Germany, 163
Gibbs, R. J., 46
Glaciers, 16(table), 21, 38(table), 52, 193
Gley, 88, 89(figure), 91
Goethite (FeOOH), 127
Goiter, 4
 Belt, 179(figure)
Gold, 185
Goldschmidt, V. M., 155
Goldschmidt-reaction principle, 97, 155
Gorsline, D. S., 120
Graphite, 135
Grassland, 157, 158(figure), 159, 160-161(table), 164(figure), 165
Gravels, 43, 44(table)
Great Basin, 21
Great Lakes, 21, 53
Great Plains, 53, 176
Green, J., 7
Greenhouse effect, 59, 196
Greenland, 21
Greenland Current, 18
Guatemala, 55
Gulf Stream, 16
Gypsum ($CaSO_4 \cdot 2H_2O$), 171, 172
Gyres, 15-16, 18

H. *See* Hydrogen
Halides, 94
Halloysite, 84
Halmyrolysates, 122
Halmyrolysis, 122
HCl. *See* Hydrochloric acid
HCO_3. *See* Bicarbonate
Helium, 52
Hemoglobin, 139
Hg. *See* Mercury
HNO_3. *See* Nitric acid
Homeostatic capacity, 4, 177
Horne, R. A., 162
H_2A, 137
H_2CO_3. *See* Carbonic acid
H_2S. *See* Hydrogen sulfide
H_2SO_4. *See* Sulfuric acid
H_3PO_4. *See* Phosphoric acid
Huang Ho, 46
Hubbard Brook region, 44, 166-167
Hudson River, 70
Human health, 4, 5, 139, 179(figure), 180-183(table). *See also* Air pollution; Atmosphere, urban; Combustion; Water pollution
Humic acid, 90, 95, 110
Humic material. *See* Humic acid; Humus
Humus, 64, 65, 88, 89(figure), 90, 95, 97, 110, 135, 171
 terrigenous, 108
Hunt, C., 20
Hurricane Agnes (1972), 48
Hydrocarbons, 52, 62
Hydrochloric acid (HCl), 62, 114
Hydrogen (H), 3(table), 94, 95, 111, 113, 114, 139, 169
Hydrogenous material, 122, 123(figure)
Hydrogen sulfide (H_2S), 52-53, 61, 137, 171
Hydrolysis curve, 69
Hydrosphere, 7(figure)
Hydroxide (OH), 174

I. *See* Iodine
Icebergs, 38(table)

Ice caps, 16(table), 21, 52, 193
Illite, 115
Inactive zone, 2, 4
India, 155
Indian Ocean, 15, 17(table), 24, 29, 30(table), 31, 61, 120, 121(figure), 193-194
Inland seas, 16(table), 21, 193
Iodine (I), 2, 3(table), 113, 177, 180-183(table)
Ions, 37, 41, 51, 65, 67, 69, 97, 104, 106, 107(table), 113, 114, 117, 125, 127, 138, 139, 142, 195. *See also* Anions; Cations; Speciation
Iron (Fe), 3(table), 41, 55, 60, 61, 67, 90, 95, 97, 99, 113, 117, 122, 125, 126, 139, 140, 141, 165, 176, 177
  concentration ratio, 142, 155
  hydroxide (Fe(OH)$_3$), 71, 84, 97, 106-107
  -manganese nodules, 126-127
  oxides, 67, 70-71, 84, 85, 90, 97
  oxyhydroxides, 84, 90, 127
  and phosphorus, 175
  -rich sulfides, 126
Irrawaddy River, 24, 26-27(table)

Japan, 18
Japan Current, 16

K. *See* Potassium
Kamchatka Current, 18
Kaolinite, 43, 84, 114, 115
Krypton, 52

La. *See* Lanthanum
Labrador Current, 18
Lagoonal interface, 24
Lakes, 16(table), 21, 53, 193
Land form, 85, 86(figure), 87(table)
Landslides. *See* Mass movement
Land total, 81
Lanthanum (La), 155
Laterites, 85, 90, 91
Leaching, 32, 65, 90, 95, 99, 126, 163, 176. *See also* Soil, surface
Lead (Pb), 3(table), 61, 67, 69, 125, 127, 180-183(table), 185
Lena River, 45
Li. *See* Lithium
Lignin, 110, 135
Likens, G. E., 162, 200
Lime, 62, 176
Limestone, 171, 176
Limonite, 90
Lithification. *See* Rock, formation
Lithium (Li), 3(table), 155, 180-183(table)
Lithogenous material. *See* Rock, -originated material
Lithosphere, 7(figure), 132(figure), 137
Litterfall, 90, 159, 160-161(table), 162, 163, 164(figure), 165, 166, 167, 171, 174
Looping, 61
Lvovitch, M. I., 200

Mackenzie, F. T., 20
Mackenzie River, 26-27(table), 29, 45
Macronutrients, 2
Magma system, 126
Magnesium (Mg), 2, 3(table), 39, 41(table), 51, 60, 61, 83, 90, 94, 95, 103, 122, 138, 139, 165, 167
Magnetohydrodynamics, 162
Malay Archipelago, 40
Manganese (Mn), 3(table), 61, 67, 69, 95, 97, 113, 117, 122, 125, 126, 139, 141, 142, 165, 177
  nodules, 113, 122, 126-127, 194
  oxide, 67, 69, 70-71, 127
  oxide, hydrous, 97
  See also Birnessite; Todorokite
Mass movement, 37, 38(table)
Mass wastage, 99
Mediterranean Sea, 15, 24, 61
Mekong River, 24, 26-27(table), 45
Mercury (Hg), 3(table), 9, 52, 60, 61, 177, 185
Metabolism, 135, 137
Metal-oxide phase, 67, 70

Metals. *See individual names*
Meteorites, 9
Methane (CH$_4$), 88, 135, 137, 169
Methyl mercury, 185
Metric conversions, 14(table note)
Mexico, Gulf of, 15
Mg. *See* Magnesium
Microorganisms, 131, 156, 163
Middle East, 54
Midocean ridge spouts, 176
Minerals, 2, 3(table), 155. *See also* Elements; *individual names*
Mississippi River, 45
Missouri River basin, 99
Mn. *See* Manganese
δMnO$_2$. *See* Birnessite
Mn$_3$O$_5$. *See* Todorokite
Mo. *See* Molybdenum
Mollisols. *See* Soil, prairie
Molybdenum (Mo), 2, 3(table), 4, 5, 127, 138, 139, 140, 142, 177, 180–183(table)
Montmorillonite, 43, 84
Moon, 9, 85
Munn, R. E., 55
Murray River, 40, 45

N. *See* Nitrogen
Na. *See* Sodium
NaCl. *See* Sodium chloride
NaSO$_4^-$. *See* Sodium sulfate anion
National Academy of Sciences, 55. *See also* Food and Nutrition Board
Na$_2$CO$_3$. *See* Sodium carbonate
Nelson River, 40
Nepheline syenite, 85
NH$_3$. *See* Ammonia
NH$_4$. *See* Ammonium
NH$_4$HSO$_4$. *See* Ammonium sulfide
(NH$_4$)$_2$SO$_4$. *See* Ammonium sulfate
Ni. *See* Nickel
Nickel (Ni), 3(table), 55, 61, 67, 111, 122, 125, 126, 127, 138, 139, 140, 176, 177, 180–183(table)
Nile River, 45, 46

Nitrate (NO$_3$), 83, 94, 113, 114, 125, 174
Nitric acid (HNO$_3$), 62, 99, 111, 173
Nitric oxide (NO), 169, 173
Nitrite (NO$_2$), 113, 173
Nitrogen (N), 2, 3(table), 41, 83, 88, 97, 111, 113, 114, 134, 135, 137, 141, 155, 156, 163, 165, 167, 169
  cycle, 173–174
  fixing, 174
  See also Free nitrogen
Nitrous oxide (N$_2$O), 137, 173
NO. *See* Nitric oxide
Nodules, 90
North America, 17(table), 23, 25(table), 34
Northern Hemisphere, 59, 120
North Polar Ocean. *See* Arctic Ocean
Northwest, 53
Norwegian Sea, 120
NO$_2$. *See* Nitrite
NO$_3$. *See* Nitrate
N$_2$. *See* Free nitrogen
N$_2$O. *See* Nitrous oxide

O. *See* Oxygen
Ocean
  biota, 110, 131–133, 134, 135, 136(table), 138, 159, 162, 172, 174, 175. *See also* Biogenic particles
  buffering system, 113, 114
  chemical composition of, 105(figure), 108(table), 109(figure), 110
  circulation, 120, 121(figure), 159, 194
  deep water, 110, 111, 120
  elements in, 103–110, 117, 123(figure), 124(table), 127(table), 138, 170, 171, 174, 175
  fluxes, 103
  geochemical processes, 103, 196, 197

Index

nepheloid layer, 117
nutrient elements in, 111–115, 120, 125, 159
particulates, 115–122, 123, 126
pH, 103, 111, 113, 114, 115
precipitates, 122, 123, 126
productivity, 159–162
reservoir, 103, 104(figure)
salinity, 103, 113
sediments, 115, 116(figure), 117 122–127, 195, 196
as sink, 137
surface, 110, 111, 115, 117, 120, 196
See also Atmosphere, oceanic; Water, oceanic; *individual names*
Odén, S., 64
OH. *See* Hydroxide
Oil, 131
Ooze, 123
Organic matter, 88, 91, 95, 107–108, 110, 111, 115, 123, 125, 135, 137, 156, 160–161(table), 169, 170, 173, 174
  phase, 67
  reservoirs, 164(figure), 165
  See also Litterfall
Orinoco River, 24, 26–27(table), 45
  basin, 34
O_2. *See* Free oxygen
Oxidation, 61, 111, 113, 122, 125, 127, 137, 139, 140
Oxidizing environment, 97–99
Oxygen (O), 3(table), 83, 111, 113, 120, 141, 169, 171
"Oxygen zero" boundary, 126

P. *See* Phosphorus
Pacific Ocean, 15, 16, 17(table), 24, 29, 30(table), 31, 61, 66, 111, 112(figure), 117, 118, 120, 121(figure), 125, 126, 127, 194.
  See also East Pacific Rise material
Paleozoic era, 140
Palladium, 185
Particulate organic material (POM), 170. *See* Organic matter; Particulates; Suspended load data
Particulates, 54–55, 59, 110, 163, 170–171, 174, 175, 194–195.
  See also Ocean, particulates
Pb. *See* Lead
Peats, 82, 167
Percolation, 90, 126, 194
Permafrost, 21
Permeability, 32
Peru Current, 16
Pesticides, 61
pH, 43, 51, 62, 64–65, 69–71, 94, 97, 99, 114, 139, 176–177, 194, 195. *See also* Ocean, pH
Phosphates (PO_4), 64, 97, 138, 175
Phosphoric acid (H_3PO_4), 111
Phosphorus (P), 2, 3(table), 41, 86, 97, 111, 114, 122, 134, 138, 141, 165, 166, 169, 177
  cycle, 174–176
Photosynthesis, 14, 81, 108, 111, 113, 114, 137, 140, 159, 162, 170, 174
Plankton, 111, 125, 162. *See also* Zooplankton
Plant
  accumulator, 155, 176
  -animal interaction, 177
  chlorosis, 4
  diseases, 4, 180–183(table)
  domination of biomass, 131, 162
  elements, 139, 140–142, 162, 165–167, 169, 174, 175
  and environment, 141, 176
  indicator, 155
  leguminous, 176
  nitrogen in, 174
  organic matter, 163, 164(figure)
  particles, 53
  and pollution effect, 142
  /soil element enrichment ratios, 142, 143–155
  trace elements in, 176–177
  See also Croplands; Forested areas; Soil, and plants; Vegetation

Plasma, 139, 142, 146–154(figure)
Platinum (Pt), 185
Podzols, 89(figure), 90, 91,
    92(figure), 95, 99
PO_4. See Phosphates
Pollutants
  human-caused. See Atmosphere,
    urban; Combustion
  metal, 53, 55
  particulates, 55
  See also Air pollution; Pesticides;
    Sediment, as pollutant; Water,
    pollutants; Water pollution
Polymers. See Carbohydrates,
    polymers
POM. See Particulate organic material
Pore water, 125
Potassium (K), 2, 3(table), 39, 51,
    60, 61, 83, 86, 90, 94, 95, 103,
    114, 138, 141, 142, 165, 166,
    167, 176, 177
Potomac River, 47
Prairie areas, 62. See also Soil,
    prairie
Precambrian era, 140
Precipitation, 13, 14, 18, 19, 23, 24,
    29, 31–32, 34, 61, 106, 193, 194
  acid, 62, 63(figure), 65, 70, 99,
    169, 173, 177, 194
  and salinity, 43
  and soil formation, 88–91
  and transport, 45, 46(figure),
    47–48, 167
  See also Water, flux
Pressure, 103
Protein, 110, 134, 138, 172
Pt. See Platinum
Pyrite (FeS_2), 172

Quartz, 84

Ra. See Radium
Radiation. See Electromagnetic
    radiation; Solar radiation
Radiogenic gases, 52
Radiolarian particles, 117

Radium (Ra), 111
Radon, 52
Rain
  chemistry, 43, 66, 137, 163
  intercepted, 31
  and leaching, 95
  ocean, 60
  and salinity, 41
  See also Canopy drop; Precipitation;
    Stem flow; Throughfall
Rainout, 61, 174
Redox
  boundary, 113
  function, 139–140
  potential, 103, 141
Regolith, 86, 88
Reservoirs, 7, 8, 9, 133, 196.
    See also Biosphere; Ocean,
    reservoir; Organic matter,
    reservoirs; Soil, reservoir;
    Water, atmospheric; Water, land
    reservoirs of; Water, oceanic
Residence time, 19–21, 22, 23, 171
  atmospheric, 61, 62
  oceanic, 29–31
Respiration, 113, 114, 137, 169, 174
Rio Grande, 39, 43
River-ocean interface. See Estuarine
    interface; Lagoonal interface
Rivers, 16(table), 23, 24, 26–27(table),
    28(figure), 33(figure), 193
  erosion rate, 117
  flow rate, 51
  and geochemical movement, 39–51,
    66–71, 115, 175
  pH, 62, 64, 69–71
  See also individual names
Rock
  -dominated areas, 41, 43
  formation, 9, 14, 122
  igneous, 43
  metamorphic, 85
  -originated material, 122, 123
  sedimentary, 135
  slides. See Mass movement
  volatilization of, 52–53

weathering of, 81, 122, 123, 172
  See also Soil horizons
Ross Sea, 120
Runoff, 16(table), 17(table), 21, 22, 23, 24, 25(table), 29, 32, 34, 39, 43, 44, 45, 90, 114, 117, 137, 167, 174, 193
  acidic, 17
  continental, 193-194
  and forests, 163
  monsoon, 45

S. *See* Sulfur
Sacramento River, 48
St. Helens, Mount, 170
St. Lawrence River, 45
Salinity. *See* Ocean, salinity; Rain, and salinity; Soil, saline; Water, salinity
San Diego (Calif.), 43
Sands, 43, 44(table), 176
Sandstone, 43, 44(table)
San Francisco Bay, 48
Sb. *See* Antimony
Sc. *See* Scandium
Scandinavia, 62
Scandinavian shield, 52
Scandium (Sc), 55, 61, 120
Se. *See* Selenium
Seas, 15. *See also* Inland seas; *individual names*
Sea spray, 54(table), 55, 60, 171, 174
Sea-surface boundary, 60
Sediment, 49-51, 61, 66-67, 71, 173, 175, 194-195
  control, 99-100
  hydrothermal, 126
  human sources of, 48
  marine, 66, 115-127
  as pollutant, 44, 49, 52
  terrigenous, 116(figure), 117
  wind-blown, 55, 61
  See also Suspended load data
Selenium (Se), 2, 3(table), 60, 61, 97, 176, 177, 180-183(table), 185

Shale, 125, 176
Shenandoah River, 47-48
Si. *See* Silicon
Silica, 90, 97, 118, 119(figure), 120
  opaline, 125
Silicate dust, 59
Silicates, 67, 99, 104, 114, 123(figure)
Silicon (Si), 3(table), 41, 111, 117, 155, 167, 177
  dioxide (SiO_2), 85, 114, 115, 117, 118
  hydroxide, 84
  oxide, 84
  oxyhydroxide, 84
Silting, 46
Silver (Ag), 3(table), 61
SiO_2. *See* Silicon, dioxide
Slope, 85
Sn. *See* Tin
Snow, 32, 62
Snowout, 61
Sodium (Na), 2, 3(table), 41, 43, 44(table), 51, 60, 83, 90, 94, 95, 103, 106, 114, 122, 138, 139, 142, 165, 167
Sodium carbonate (Na_2CO_3), 106
Sodium chloride (NaCl), 43
Sodium sulfate anion ($NaSO_4^-$), 106
Soil
  air, 82-83
  and animals, 86, 88
  arable, 81, 94. *See also* Croplands
  and climate, 88-91
  composition, 82-83, 84(table), 85, 91-95
  creep. *See* Mass movement
  -crust enrichment ratios, 55, 58(table), 60, 144-145(table)
  desert, 89(figure), 91, 92(figure)
  element movement in, 95, 97
  elements in, 84-85, 91, 93(figure), 94-99, 138, 165-166, 171, 175, 176
  exhaustion, 81-82
  formation, 83-91
  genetic, 83, 188

Soil, *Cont.*
  histic, 88
  horizons, 83, 88, 90
  infiltrability, 32
  muck, 82
  and plants, 85–86, 88, 97, 141–155, 164(figure)
  prairie, 89(figure), 90–91, 92(figure), 95
  renewal, 81, 82
  reservoir, 81–83
  saline, 83, 94
  sub-, 83, 85, 90, 91, 95, 99, 194
  surface, 83, 85, 90
  tropical, 90, 92(figure), 99, 166
  unweathered rock, 83, 99
  water, 21–22, 32, 38(table), 83, 176, 177, 193, 194
  weathered parent rock, 83, 85, 86, 163
  weathering, 84, 85, 97, 100
  See also Atmosphere, and soil interface; Erosion, soil; Podzols
Solar energy, 193
Solar radiation, 59, 193, 196
Solubility. *See* Dissolved load data; Elements, solubility
Solum. *See* Soil, genetic
Sorption, 69–71
SO_2. *See* Sulfur dioxide
SO_4. *See* Sulfate
South America, 17(table), 23, 24, 25(table), 34, 40
Southern Hemisphere, 59
South Polar region. *See* Antarctica
Soviet Union, 40, 155
Spanish Moss, 53, 142
Speciation, 106–107, 108(table), 109(figure)
Spodosols. *See* Podzols
Spruce trees, 31
Sr. *See* Strontium
$SrSO_4$. *See* Strontium sulfate
Starch, 135
Stem flow, 31, 38(table), 167
Steppes, 167, 168(figure)
Stones, 139

Stratosphere, 61, 196
Streams, 16(table), 21, 22, 114, 115, 193
  and element transport, 37, 39, 44, 87(table), 167, 172, 175, 194–195. *See also* Rivers, and geochemical movement
Strontium (Sr), 3(table), 5, 43, 60, 61, 111, 165, 176
Strontium sulfate ($SrSO_4$), 111
Sulfate (SO_4), 41(table), 51, 60, 62, 64, 65, 83, 94, 97, 106, 125, 126, 138, 171, 172, 174
Sulfides, 67, 122, 126, 137
Sulfur (S), 2, 3(table), 55, 59, 65, 97, 103, 113, 126, 134, 137, 142, 169, 177, 185
  cycle, 171–173
Sulfur dioxide (SO_2), 39, 61, 114, 137, 169, 171, 172, 194
Sulfuric acid (H_2SO_4), 59, 62, 99, 172, 173
Sunlight, 159, 162
Surface charge. *See* Anions; Cations
Suspended load data, 44–49, 50(figure), 194. *See also* Colloids
Susquehanna River, 48, 49, 50
Sweden, 64

Te. *See* Tellurium
Tellurium (Te), 3(table), 177, 180–183(table), 185
Temperate zone, 156, 159, 163, 164(figure), 165, 166, 167, 168(figure)
Temperature, 103, 113. *See also* Biomes; Climate; Greenhouse effect
Terpenes, 53
Th. *See* Thorium
Thallium (Tl), 69, 185
Thorium (Th), 3(table), 61, 120, 155
Throughfall, 31, 163
Thunderstorms, 163
Ti. *See* Titanium
Tidewater, 46–47

Tin (Sn), 2, 3(table), 53, 185
Titanium (Ti), 3(table), 120, 142, 155
Tl. *See* Thallium
Todorokite (Mn_3O_5), 127
Topography, 45, 46(figure)
Total particulate matter (TPM), 115, 117
Toxicity, 4, 61, 179, 180-183(table), 185, 195
Toynbee, A. J., 199
TPM. *See* Total particulate matter
Trace elements, 2, 3(table), 4, 5, 58(table), 69, 95, 96(table), 106, 108(table), 111, 117, 120, 124(table), 125, 127, 139-141, 176-185, 197
 and disease, 179, 180-183(table)
 vapor, 179, 185
 "veil" theory, 126
Trade winds, 19, 61, 65, 195
Transpiration, 13, 14, 23, 32, 34, 53
Transportation, 9, 14, 67, 126. *See also* Geochemical ecology, movement
Tree sap, 165
Tropical zone, 156, 157-159, 162, 163, 164(figure), 165, 167, 196-197. *See also* Soil, tropical
Tropopause, 61, 62, 66
Troposphere, 61, 196
Tundra, 156, 157, 158(figure), 160-161(table), 164(figure), 165, 167, 168(figure)
Turekian, K. T., 5, 106

UNESCO. *See* United Nations Educational, Scientific and Cultural Organization
United Nations Educational, Scientific and Cultural Organization (UNESCO), 34
United States
 acid rain in, 62, 63(figure), 99
 mineral exploration, 155
 soil, 91, 92(figure), 176
 trace elements, 178-179(figure)
 wind erosion, 54, 56-57(figure)

Uplift, 14
Uranium, 52
Urey, H. C., 114

V. *See* Vanadium
Vadose water, 21, 22
Value ranges, 8-9, 41(table)
Vanadate anion (VO_4^-), 97
Vanadium (V), 3(table), 138, 140, 180-183(table)
Vegetation, 34, 49, 53, 67, 90, 171
 cycle, 22, 23
 See also Rain, intercepted; Soil, and plants; Transpiration; Water, and plants
Venus, 9, 59
Vermiculite, 84, 99
Vertical mixing, 195, 196
VO_4^-. *See* Vanadate anion
Volatiles, 31
Volatilization, 52, 65, 172
Volcanoes, 54(table), 55, 58(table), 59, 122, 126, 137, 170, 171, 196
Volga River, 45

Wales, 67, 185
Washout, 61
Water
 atmospheric, 17(table), 18-19, 31-32, 34, 193, 194
 and biota, 23, 31-32, 34, 67, 70, 134, 135(figure), 193, 194
 buffering capacity of, 64, 69
 chemistry, 42(figure), 43-44
 consumption, 14-15
 cycle, 13-14, 17(table)
 as electrolyte, 69
 erosion, 9, 14, 43, 47-49, 54, 117
 flow, 22-23, 31, 32
 flux, 16(table), 17(table), 19-21, 23-34, 60, 193-195
 ground, 13-14, 21, 22-23, 38(table), 43-44, 62, 64, 193
 land reservoirs of, 16(table), 17(table), 21-23
 motion. *See* Gyres

Water, *Cont.*
  oceanic, 15-18, 20, 29-31, 193, 194
  and oxidation of organic matter, 111
  and plants, 31-32, 34
  pollutants, 15, 23, 24, 31, 44.
    See also Water pollution
  quality, 200
  salinity, 18, 24, 37, 39, 40, 41,
    42(figure), 43, 44(table), 49, 115,
    195
  surface. *See* Rivers; Streams
  underground, 21-22, 32
  uses, 14
  See also Geochemical ecology,
    movement; pH; Soil, water
Water pollution, 179, 200
Weathering. *See* Chemical weathering
Weddell Sea, 120
Westerlies, 19, 65
Whelpdale, D. M., 55

Whittaker, R. H., 162, 200
Wind carrying power, 18-19, 37.
  See also Atmosphere, and
  geochemical movement; Erosion,
  wind; Trade winds; Westerlies

Xenon, 52

Yangtze River, 24, 26-27(table)
Yukon River, 46, 68(figure)

Zambezi River, 45
Zinc (Zn), 3(table), 55, 60, 61,
  67, 69, 125, 126, 127, 139,
  141, 142, 176, 177,
  180-183(table)
Zirconium (Zr), 3(table), 155
Zn. *See* Zinc
Zooplankton, 117
Zr. *See* Zirconium